HDTV
FOR
DUMMIES®

by Danny Briere and Pat Hurley

WILEY

Wiley Publishing, Inc.

HDTV For Dummies®

Published by
Wiley Publishing, Inc.
111 River Street
Hoboken, NJ 07030-5774

Copyright © 2005 by Wiley Publishing, Inc., Indianapolis, Indiana

Published by Wiley Publishing, Inc., Indianapolis, Indiana

Published simultaneously in Canada

For general information on our other products and services, please contact our Customer Care Department within the U.S. at 800-762-2974, outside the U.S. at 317-572-3993, or fax 317-572-4002.

For technical support, please visit www.wiley.com/techsupport.

Wiley also publishes its books in a variety of electronic formats. Some content that appears in print may not be available in electronic books.

Library of Congress Control Number: 2004107890

ISBN: 0-7645-7586-4

Manufactured in the United States of America

10 9 8 7 6 5 4 3 2 1

1B/QT/RR/QU/IN

WILEY

About the Authors

Danny Briere founded TeleChoice, Inc., a telecommunications consulting company, in 1985 and now serves as CEO of the company. Widely known throughout the telecommunications and networking industry, Danny has written more than one thousand articles about telecommunications topics and has authored or edited ten books, including *Internet Telephony For Dummies*, *Smart Homes For Dummies* (now in its second edition), *Wireless Home Networking For Dummies*, *Windows XP Media Center Edition 2004 PC For Dummies*, and *Home Theater For Dummies*. He is frequently quoted by leading publications on telecommunications and technology topics and can often be seen on major TV networks, providing analysis on the latest communications news and breakthroughs. Danny splits his time between Mansfield Center, Connecticut, and his island home on Great Diamond Island, ME, with his wife and four children.

Pat Hurley is Director of research with TeleChoice, Inc. specializing in emerging telecommunications and digital home technologies, particularly all the latest consumer electronics, access gear, and home technologies, including wireless LANs, DSL, cable modems, satellite services, and home-networking services. Pat frequently consults with the leading telecommunications carriers, equipment vendors, consumer goods manufacturers, and other players in the telecommunications and consumer electronics industries. Pat is the coauthor of *Internet Telephony For Dummies*, *Smart Homes For Dummies*, *Wireless Home Networking For Dummies*, *Windows XP Media Center Edition 2004 PC For Dummies*, and *Home Theater For Dummies*. He lives in San Diego, California, with his wife, new baby girl named Annabel, and two smelly dogs.

Authors' Acknowledgments

Danny wants to thank his wife for not freaking out when he's gotten big TV sets and lots of equipment, and for accepting his bribes of absolutely everything he could think of to buy the extra time to research this book. She's a trooper. However, this "Thank You" is not about her, but about co-author Pat, who's been so much fun to work with on these *For Dummies* books over the years. While Pat lives on the other coast of the U.S., and we rarely actually see each other, we work side by side through videoconferencing, IM, and e-mail, and (oh, yes) voice calls. I've heard his new child's voice and wife's pleas for dinner in the background, and feel I know his house by the location of the different devices we so often write about ("I'm IMing you from the Media Center PC in the living room . . ."). Pat's wife is used to Danny calls after hours and on weekends, and has been nice enough to let those slip by — well at least occasionally — and Danny thanks her immensely for never chewing him out on the phone (she chews out Pat instead, I'm sure, but Pat's too Navy to tell his superior about that). When Pat went in the hospital in the early stages of the book, it simply was not as much fun writing a book without Pat's incessant rantings about how stupid people are in the marketing of their products, and about how nobody's Web site is worth a hoot (being a good former Navy man, he actually never says *hoot*). These books are always a pain in the ruckus to do, but are made all the more pleasant with a co-writer who has a sense of humor (or the smarts to laugh at his boss's bad jokes, whichever applies).

Pat dedicates this book to his beautiful little girl, Annabel Stone Hurley. You're never too young to enjoy a little TV, Annabel! Now if they would only start broadcasting Sesame Street in HD! Pat also profusely thanks his wife Christine, without whom none of this good stuff (babies, books, and the like) would be possible. Pat also must give a nod to Dr. Bard Cosman, surgeon *par excellence* at the U.C. San Diego Medical Center. Thanks, Dr. Cosman, for getting me back on my feet to finish this book!

We would also like to thank Jenny Miller at the Consumer Electronics Association (CEA), for her assistance and guidance.

Publisher's Acknowledgments

We're proud of this book; please send us your comments through our online registration form located at www.dummies.com/register/.

Some of the people who helped bring this book to market include the following:

Acquisitions, Editorial, and Media Development

Project Editor: Pat O'Brien

Acquisitions Editor: Melody Layne

Senior Copy Editor: Barry Childs-Helton

Technical Editor: Dale Cripps, *HDTV Magazine*

Editorial Manager: Kevin Kirschner

Media Development Supervisor: Richard Graves

Editorial Assistant: Amanda Foxworth

Cartoons: Rich Tennant (www.the5thwave.com)

Production

Project Coordinator: Emily Wichlinski

Layout and Graphics: Andrea Dahl, Lauren Goddard, Joyce Haughey, Stephanie D. Jumper, Jacque Roth, Heather Ryan, Rashell Smith

Proofreaders: Joe Niesen, Carl William Pierce, Rob Springer, TECHBOOKS Production Services

Indexer: TECHBOOKS Production Services

Publishing and Editorial for Technology Dummies

> **Richard Swadley,** Vice President and Executive Group Publisher
>
> **Andy Cummings,** Vice President and Publisher
>
> **Mary Bednarek,** Executive Acquisitions Director
>
> **Mary C. Corder,** Editorial Director

Publishing for Consumer Dummies

> **Diane Graves Steele,** Vice President and Publisher
>
> **Joyce Pepple,** Acquisitions Director

Composition Services

> **Gerry Fahey,** Vice President of Production Services
>
> **Debbie Stailey,** Director of Composition Services

Contents at a Glance

Table of Contents

Introduction

*W*elcome to *HDTV For Dummies*. HDTV is the hottest technology to hit your local electronics store since the advent of cell phones. HDTVs are getting bigger, better, cheaper, more sophisticated, and more useful every day. Since you've bought this book, we figure that not only do you agree with us, you're already part of the HDTV movement. To arms, Comrade!

One of the most appealing things about the current crop of HDTVs is the ease with which you can set up an HDTV-powered home theater, including surround sound and an awesome picture. However, the rapidly dropping price of HDTVs might be the most attractive aspect of all — we've reached a point in time where you don't have to be rich to consider an HDTV for the bedroom, too! Figuring out which HDTV to buy can be confusing, as there are all sorts of technologies, sizes, standards, etc. And then, once you've decided on what to buy, making sure it will work with all your other gear — your DVD, camcorder, VCR, set-top box, and so on — can be even more confusing. That's where this book comes in handy. Our aim is to make sure you get the most bang for your buck (or franc, or peso, or whatever — even euros!).

About This Book

If you're thinking of purchasing an HDTV and installing it in your home, this is the book for you. Even if you've already purchased the HDTV itself, this book will help you install and configure the HDTV. What's more, this book helps you get the most out of your investment after it's up and running.

With this book in hand, you'll have all the information that you need to know about the following topics:

- Planning your HDTV system, including all sorts of accessories
- Evaluating and selecting the right HDTV for your home
- Installing and configuring the HDTV equipment in your home
- Hooking up your HDTV to the right high-definition programming sources
- Adding A/V entertainment gear and accessories to your HDTV

- Playing video games on your Xbox, in high-def splendor

- Accessing your HDTV from around the house over a home network

- Enhancing your HDTV environment so you can have your own HDTV theater

Foolish Assumptions

While writing this book, we had the perfect excuse to watch a lot of TV. ("Not now, Honey, I'm working on my book.") It's been great.

Still, we already know what we like and what we value in an HDTV. But we're writing this book for you! While writing, we ponder all kinds of questions concerning our readers. Who are you? Where are you? What did you eat for lunch? Which movies tweak your interest? How do your HDTV desires line up with your budget? Queries like that fill our minds constantly, much to the consternation of our spouses, who prefer more useful thoughts like "Shouldn't I take the trash out, or empty the Diaper Genie?"

Because we never get to meet you in person, we end up making a few assumptions about you and what you want from this book. Here's a peek at our thoughts about you:

- You love movies, television shows, or video games — or perhaps all three.

- You've experienced wide screens and surround sound at the theater, and you liked it.

- For one reason or another, a 19-inch TV set with a single built-in speaker doesn't adequately meet your audio or video entertainment needs.

- You probably own a computer, or will soon.

- You don't shy away from high-tech products, but you also aren't the first person on the block with the latest electronic goodie.

- The weird technicalities of your A/V system make you dizzier than a Marilyn Monroe movie.

- You know something about the Internet and the Web.

- You, or someone in your family, enjoy watching movies, listening to MP3 audio, playing games, and possibly making movies on your computer.

If that describes you today, in a prior life, or in another personality, then this book is for you.

How This Book Is Organized

This book is organized into several chapters that are grouped into eight parts. The chapters are presented in a logical order — flowing from purchasing your HDTV, through all the things you'd want to hook up to it to exploit its very existence, to some detailed drill-down discussions about high-definition topics that will help you get the most out of your HDTV environment. However, you can feel free to use the book as a reference and read the chapters in any order that you want. We wrote it that way.

Part 1: HDTV Fundamentals

The first part of the book is a primer on HDTV. If you've never owned an advanced level television — much less attempted to install one — this part of the book provides all the background information and techno-geek lingo that you need to feel comfortable. Chapter 1 presents general HDTV concepts; Chapter 2 discusses the most popular HDTV technologies and familiarizes you with high-definition terminology, and also provides guidance on making buying decisions; and Chapter 3 introduces you to several popular ports, interfaces, jacks, plugs, cables and the sort — everything you'll need to know about connecting your HDTV into your existing audio/visual environment.

Part II: Love at First Sight

The second part of the book helps you install your HDTV system. It helps you decide what you will be connecting to the HDTV, how to enhance your HDTV so that it is tuned perfectly for your use, and also tells you about a range of little "black boxes" that can help you optimize all of the *non-HDTV* signals you send to your HDTV, making them look better on your HDTV big screen.

Part III: HDTV Channels

Part III discusses all the different forms of high-definition signals that you can access and/or subscribe to, in order to really take advantage of your great new investment. In the first chapter of this part, we talk at a high level about what is available in high-definition format now, and what's coming in the near future. Then in the next three chapters, we dive into each option — over-the-air broadcasts, cable, and satellite — so you know where you can get what signals, and which is best (at least in our eyes).

Part IV: Movie Machines

The broadcast programming discussed in Part III is nice, but let's face it, we all want to watch a lot of other content, too — we're talking about all those DVDs and VHS movies you own. In this part, we talk about the complexities of interfacing your DVD player/recorders and VHS VCRs with your HDTV. We also delve into the exciting world of digital video recorders — DVRs or TiVos as some people call them (referring to one brand on the market generically). With these devices you can record all sorts of content for later watching.

Part V: Monitor Madness

After you get your HDTV system installed and running, you will certainly want to use it for even neater things if you can. Part V of the book presents many cool things that you can do with your HDTV, including playing multi-user computer games, connecting your camcorder to preview your future *America's Funniest Home Videos* submission, and operating various types of smart home conveniences from the luxury of your bedroom. This part also describes how to use a home network to connect your HDTV to other parts of the house and to the Internet — all with the intent of making your HDTV investment simply more accessible.

Part VI: Sensory Overload

In this part, we spend a lot of time drilling down in a lot of detail about the nitty-gritty of getting your home HDTV-viewing experience as good as you can get it. We start with an extensive discussion about audio basics and how they affect your HDTV viewing experience. We then discuss built-in speaker options versus external sound systems, and the advantages of a surround-sound-powered HDTV experience. We'll also tell you about how to use lighting, room treatments, and other nuances in your home to create a true HDTV theater. Finally, we go into a series of chapters that delve down into the details of your TV picture, and the various ways that HDTV can be accomplished, including front and rear projection, plasma and LCD screens, and the good old CRT approach. When you've finished this part, you should know *much* more than the average salesperson walking the show floor at your local TV store.

Part VII: Geek Stuff

Of course, the more you know, the geekier these topics start to appear. Before long, you may be dreaming of ultra-high-tech ways to expand your system's capabilities. This part of the book unashamedly encourages that bad habit.

Part VIII: The Part of Tens

Part VIII provides three top-ten lists that we think you'll find interesting — ten places to look online and locally to buy an HDTV; ten devices to connect to your HDTV; and ten frequently asked questions about HDTV.

Icons Used in This Book

All of us these days are hyper-busy people, with no time to waste. To help you find the especially useful nuggets of information in this book, we've marked the information with little icons in the margin. The following icons are used in this book:

As you can probably guess, the Tip icon calls your attention to information that will save you time or maybe even money. If your time is really crunched, you might try just skimming through the book and reading the tips.

The little bomb in the margin should alert you to pay close attention and tread softly. You don't want to waste time or money fixing a problem that could have been avoided in the first place.

This icon is your clue that you should take special note of the advice that you find there . . . or that this paragraph reinforces information that has been provided elsewhere in the book. Bottom line: You will accomplish the task more effectively if you remember this information.

Face it, HDTVs and home entertainment systems are high-tech toys that make use of some pretty complicated technology. For the most part, however, you don't need to know how it all works. The Technical Stuff icon identifies the paragraphs that you can simply skip if you're in a hurry or you just don't care to know.

Where to Go from Here

Where you should go next in this book depends on where you are in the process of planning, buying, installing, configuring, and/or using your HDTV. If HDTV in particular is totally new to you, we recommend that you start at the beginning with Part I. If you feel comfortable with HDTVs and all of its connections, you might just read Chapter 2 about buying advice. If you've already got your HDTV, you might want to check out how to make sure it's optimally configured with the chapters in Part II. If your HDTV is installed and you want to know more about what you can connect to it, Parts III, IV, and V all talk about neat things you can channel (oops, pun) to your HDTV. If you are in the depths of analyzing your equipment options, Part VI might be the best place to find the details you are looking for. There's simply a lot here for whatever you need to know about HDTV.

Or, if you're into Fate, you can just open the book to any page and start reading.

Either way, happy reading!

Part I
HDTV Fundamentals

The 5th Wave   By Rich Tennant

RICHTENNANT

"HD TV, huh? Well'p there goes the ambiance."

In this part . . .

*I*f you ever had the exciting opportunity to go to FAO Schwarz's flagship store in New York City before it closed, you were greeted by a huge fanciful clock, made even more famous in the movie *Big*. The song played by the clock is very appropriate here as we begin to talk about HDTVs — a chiming "Welcome to our world, welcome to our world, welcome to our world of toys!"

Oh boy, are HDTVs fun — and that fun, in Part I, is just beginning. We're going to introduce you to the world of HDTV, our world of toys. In Chapter 1, we'll introduce you to the key acronym of the HDTV world, ATSC, and tell you why you should care about it. We'll explain the foundations of HDTV, of things like resolution, scan types, and aspect ratios. We'll talk of 480i, 480p, 720p and 1080i, which sound more like something from *I, Robot* than from Circuit City. Don't worry — we explain what all that means!

We'll also help you go shopping. In Chapter 2, we talk of the key buying criteria for HDTVs and how to best match a TV to your needs, environment, existing audio/video gear, and other HDTV decision-affecting facets of your life.

And then we'll wrap up our introduction by making sure your basis in HDTV technology is sound by talking about the backs, sides, tops, and other parts of the HDTV systems — all the places where you'd connect to your HDTV display other sources and gear, like DVDs, VCRs, camcorders, satellite receivers, video jukeboxes, and yes, even bathtubs.

You're off on your HDTV adventure if you're starting with Part I right away. Be sure to turn off your toys when you are done playing with them!

Chapter 1

What the Heck Is HDTV?

Since the transition to color TV in the 1950s and '60s, nothing — nothing!! — has had as much impact on the TV world as HDTV (high-definition TV) and digital TV. That's right, TV is going digital, following in the footsteps of, well, everything.

We're in the early days of this transition — a lot of TV programming is still all-analog, for example — and this stage of the game can be confusing. In this chapter, we alleviate HDTV anxiety by telling you what you need to know about HDTV, ATSC, DTV, and a bunch of other acronyms and tech terms. We also tell you *why* you'd want to know these terms and concepts — how great HDTV is, and what an improvement it is over today's analog TV (as you'll see when you tune in to HDTV). Finally, we guide you through the confusing back alleys of HDTV and digital TV — making sure you know what's HDTV and what's not.

Oh, Say, Can You ATSC?

A long time ago (over 50 years ago — longer than even Danny has been alive!), in a galaxy far, far . . . errr, actually right here in the U.S. . . . a group called the *NTSC* (National Television System Committee) put together a group of technical specifications and standards that define television as we know it today. Sure, there have been some changes in those 50 years (such as the addition of color), but today's analog TVs are built on this NTSC system.

Fifty years is a long time for any technology to dominate. Indeed, technologies and components used in television-transmission systems, cameras, recording systems, and display systems (the TVs themselves) have long been capable of doing something more.

In the 1980s, the *ATSC* (Advanced Television System Committee) was formed to move TV forward. Many years later (1996), the ATSC's recommendations for a digital-television system were adopted by the FCC (Federal Communications Commission — the folks who set standards for TV broadcasts, regulate phone companies, and fine Howard Stern). ATSC standards use newer-than-1953 technology to give you TV like you've never had before:

- ✔ Widescreen images like those in the movies
- ✔ Greater detail — up to six times more detail
- ✔ Sharper images
- ✔ Smoother, more filmlike images with no video flicker
- ✔ All digital, with none of the "ghosts" and other image problems found in analog TV

Powerful Performance

HDTV (and digital TV, DTV, in general — there are some digital TV variants that are *not* high-definition, and we discuss them in this section) is all about giving you a bigger and better picture, better audio, and generally making your TV-watching experience more like a movie-watching experience. In fact, at its best, HDTV is so realistic that it's often described as "looking through a window" — as if you're really there, not just watching a program.

Video standards

There are three essential concepts to understand when you are comparing different video standards:

- ✔ **Resolution:** the number of individual picture elements that make up a TV image. The higher the resolution, the more detailed the image, and the sharper the image will appear.

  Resolution is defined by one of two factors:

  - *Lines* (the number of left-to-right lines the TV can display). CRT-based TVs (tube TVs) are rated this way.
  - *Pixels* (the number of pixels across the screen times the number up and down). Fixed-pixel displays (plasmas, LCDs, DLPs and the like) are rated this way.

✔ **Scan Type** comes in two forms:

- *Interlaced scan:* These TV images are created by lighting up every other row of horizontal lines on the screen in one instant, and then going back through and lighting up the remainder of the lines in the next instant. It happens so fast that your eye can't really tell it's happening.

- *Progressive scan:* These systems light all the horizontal lines in the same instant, which can make the image seem "smoother" and more like film (or real life).

✔ **Aspect Ratio** (the *shape* of your TV picture):

- Traditional TVs have a 4:3 *aspect ratio* (screen shape). This means that for every 4 units of measure across the screen, you have 3 units of screen height. For example, if the screen is 12 inches wide, it will be 9 inches high.

- HDTVs have a 16:9 aspect ratio — which makes the screen relatively much wider for the same height, compared to a 4:3 TV. Most movies are widescreen (16:9, or even wider), so HDTVs can display most movies without the annoying "letterbox" black bars on the top and bottom of the screen. Figure 1-1 compares aspect ratios.

4:3 / 1.33:1 Standard TV and older movies	16:9 / 1.78:1 US Digital TV (HDTV)

Figure 1-1: Going widescreen with a 16:9 aspect ratio.

We don't get bogged down in up-front technical explanations of these concepts. If you want to know all there is to know about such TV concepts as resolution, pixels, and interlacing, run (don't walk) to Chapter 21 right now. We'll still be here when you come back.

HDTV standards

There isn't a single "HDTV" standard out there. Instead, ATSC contains many different TV standards (with different resolutions, aspect ratios, and scan types) — 18, in fact. Some of these standards are truly HDTV; most are not. In the real world, you will deal with four standards when you try to watch TV content on your HDTV. The two primary HDTV standards are these:

✔ **720p:** This provides 720 lines of resolution with progressive scan (hence the *p*). By comparison, NTSC has less than 480 lines of resolution. 720p uses a 16:9, a widescreen aspect ratio.

✔ **1080i:** This variant (the highest resolution within the ATSC standard) uses interlaced scanning, but provides 1080 lines of resolution. 1080i is also widescreen, with a 16:9 aspect ratio.

There is actually a higher HDTV variant in the ATSC standard — 1080p, which is a progressive scan variant of 1080i. Only a few HDTV projectors (in the $40,000 and above price range) can handle this variant, and we know of *no* material that is broadcast or otherwise available as 1080p. So don't worry about it.

True HDTV performance requires at least 720p performance. If a TV program, movie, or other content is not at least 720p (either 720p or 1080i), it is *not* HDTV. If a TV can't display at least 720 lines of resolution, it is *not* HDTV-capable.

If a salesperson tries to tell you that an inexpensive plasma set, regular DVD, regular digital cable, or regular satellite TV "is" HDTV just because it's *digital,* it's not so.

Compatible DTV standards

720p and 1080i are the two HDTV standards, but you'll also find a lot of digital TV material will be broadcast at lower resolutions that don't quite make the grade as HDTV. You can still watch this programming on your HDTV — in fact, most HDTVs will make this programming look better than it does on a regular TV — but remember: That stuff *is not* really HDTV.

✔ **480p (EDTV):** This *enhanced-definition* TV standard provides higher-than-NTSC resolution, with progressive scan (NTSC is interlaced). EDTV can be (and often is) 16:9 widescreen, but it is not required to be widescreen.

✔ **480i (SDTV):** This is interlaced, non-widescreen (4:3), standard-definition TV, equivalent to NTSC analog broadcasts.

Remember these different terms — HDTV, EDTV, and SDTV — when shopping. They will often be in the product descriptions; you need to know exactly what you are buying.

Audio standards

The ATSC standard includes big improvements in the audio part of television — what you hear as part of any movie, video, or TV show. That's because ATSC includes Dolby Digital surround sound capability in the overall standard for digital TV.

Dolby Digital (which we discuss in greater detail in Chapter 18) doesn't *always* mean surround sound. Some Dolby Digital sound-tracks are stereo (two channels) or even mono (one channel). ATSC supports surround sound if a program's producer and broadcaster want to include it.

The NTSC broadcast standard supports only stereo audio (two channels) and not surround sound. Luckily, most DVDs (and some satellite and digital cable TV channels) include Dolby Digital soundtracks that can provide true surround sound. You can also use a home-theater receiver that supports systems like Dolby Pro Logic II (see Chapter 18) to create surround sound from these sources.

Dolby Digital, and surround sound in general, provides an audio soundtrack for TV shows and movies that — wait for it! — *surrounds* you and provides audio that matches the action on-screen. For example, surround sound might use speakers mounted in the rear of the room to reproduce ambient noises of the setting around the action, or give a 3D sense of space to those creepy footfalls of the bad guy sneaking up behind the protagonist.

Dolby Digital provides six channels (confusingly called *5.1)* of audio. Here's what they do:

- A center channel carries the dialogue being spoken by characters on your HDTV screen.

- Two main front channels handle left and right sound cues (and the soundtrack music) in stereo.

- Two surround channels (mounted in the rear of the room, as described earlier) provide a sense of 3D space.

- A Low-Frequency Effects (LFE) channel conveys deep bass sounds (such as exhausts rumbling and bombs exploding). The LFE channel is the ".1" in the 5.1 naming scheme for Dolby Digital. It doesn't get a whole number because it contains only low-frequency sounds, not the full range of human hearing.

Figure 1-2 shows a typical Dolby Digital surround-sound layout.

We talk about surround sound in much more detail — including details on what sort of equipment you need to hear it properly in your HDTV viewing room — in Chapter 19.

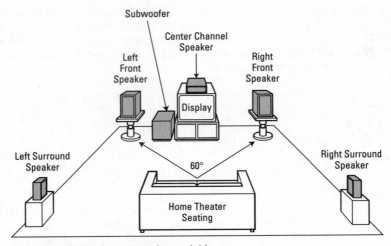

Figure 1-2: Doing the surround-sound thing.

Perplexing Pitfalls

HDTV isn't the easiest thing in the world to get figured out — we've been dealing with it for years and still run into advertising and marketing mumbo-jumbo that make us say, "Huh???" The whole purpose of *HDTV For Dummies* is to help you wade through the marketing manure and to get you up to speed on HDTV. So without further ado, here's a list of HDTV danger zones:

- ✔ **Digital confusion:** The biggest (and most prevalent) myth we see in the HDTV world is the notion that any kind of digital TV signal (such as digital cable, digital satellite, or DVD) is HDTV. This simply isn't true — a TV signal must be 720p resolution or higher to be considered high-definition.

- ✔ **EDTV confusion:** EDTVs are TVs (typically plasma flat-panel models) that cost a lot and can display progressive-scan images — but don't meet the minimum requirement of 720p, so they don't display true HDTV signals. There's nothing wrong with EDTVs, just don't be fooled into thinking you're getting an HDTV when you're not.

- ✔ **Image scaling:** We're starting to see some new marketing being applied to an old concept — *image scalers* that can convert video signals from one resolution to another.

These devices are now being marketed as "HDTV upscalers" (yeah right), with a promise that they make any TV signal into HDTV. Don't believe it. Image scalers *can* improve SDTV and NTSC images with an HDTV, but they don't make those images *into* HDTV images.

✔ **The DTV tuner:** As HDTV (and DTV in general) becomes more prevalent, *DTV tuners* will become common. These tuners (discussed in Chapter 8) let older TVs "watch" DTV broadcasts. DTV tuners do *not* turn older analog TVs into HDTVs. They just convert DTV signals to NTSC for display on an analog TV.

Chapter 2

Shopping Smart

*W*e've all been there — you're standing in the electronics store looking at a wall of TVs, all tuned to the same channel, and they all pretty much look the same. So many TV sets, so little time, so hard to choose. So you pick the one on sale and leave, happy that you got "a deal." Been there, done that.

But no more. Now we're more educated. We KNOW that those TV sets are all misconfigured to appear a certain way in the bright lights of an electronics show floor. We know to check how many digital interfaces the box has, and how deep the chassis is, and how . . . well, lots of "hows."

Choosing the right HDTV for you is not the easiest thing to do. Heck, we wrote the book on it and we still argue with each other about which HDTVs have the best bang for the buck. It's going to depend on what you are trying to do, how much money you have, and what other A/V gear you have or intend to buy.

In this chapter, we're going to walk you through a veritable Buyer's Guide to HDTVs — what to look for when shopping for just the right HDTV set for you. You can have too much HDTV (believe it or not) and the wrong type of TV for your intended use. Before you plunk down a lot of money on your well-earned HDTV surprise, make sure you're the best-informed buyer out there. Read on.

The 50,000-Foot View of HDTV

When you're trying to pick out the right HDTV for your needs, the available products break down into three major product groups, distinguished from each other by their display technology and cabinet type. By comparing display technology and cabinet type to your needs, you can easily rule out a whole bunch of different TVs, and home in on the likely best ones for you.

HDTVs come in all sorts of different sizes and shapes. Some are flat-panels that you can hang on the wall; others are projection systems much like what you'd find in a movie theater. And, of course, there are HDTVs based on tubes that look just the way TVs have for decades (only with a better picture).

Each form of HDTV has some advantages and disadvantages. In Chapters 21 through 24, we discuss these pros and cons in much more detail — here we want to give you just a quick overview to help you on your way. Skip ahead if you need the details and supporting information.

Flat-panel HDTVs

Flat-panel TVs — the super-thin models that you can literally hang on the wall — are the sexiest HDTVs available. These are the ones you see on *MTV Cribs* and that you might install in your tricked-out Escalade (yeah right). They're also good HDTVs. There are two main display technologies for flat-panel HDTVs:

- **Plasma:** These are the biggest flat-screens available, using a layer of gas trapped between two glass screens to create their images.

    - **Pros:** thin, sexy, good picture, good color

    - **Cons:** not all are HDTV, less-than-perfect black, screen burn-in, costly

    You'll see us mention "blacks" here. We're talking about how well an HDTV screen can reproduce dark tones and scenes on-screen — how well it creates black rather than gray colors.

- **LCD:** These flat-panel TVs use liquid crystal displays, just like those used in laptop computers.

    - **Pros:** same as plasma, plus no burn-in

    - **Cons:** black is poorest, costly, angle of view

Projection HDTVs

These are the TVs that project their picture from a smaller image source (either three small picture tubes, or a digital system known as a *microprojector*) onto a screen. The screen can be either part of the HDTV itself (rear projection) or a separate screen hung on your wall (front projection).

✔ **Front-Projection HDTVs:** These are the HDTV equivalents to movie theater projectors, with a big screen on the wall, and a separate projector mounted somewhere across the room.

  - **Pros:** biggest screen, potentially best picture

  - **Con:** expensive, complicated, requires setup/focus/maintenance

✔ **Rear projection HDTVs:** The picture is projected on the back of a screen that is built into the HDTV itself.

  - **Pros:** best bargain, no burn-in with microprojectors, near flat-panel thinness for microprojector

  - **Cons:** burn-in for CRT, expense for microprojector, size for CRT

CRT HDTVs

The final category of HDTVs is based on the good old-fashioned picture tube — also known as the CRT, or cathode-ray tube.

✔ **Pros:** cheapest, great color, great blacks

✔ **Cons:** smallest screen, bulky, lower resolution than digital displays

What's Important in an HDTV

When looking at HDTVs, we think the following are the most important buying criteria for your purchase:

✔ **What's your budget?** We don't mean just for the TV set, but also for any attached home theater surround-sound system, special remote controls, automated drapes, lighting controls, popcorn poppers, and the like. It makes a big difference if you are building a home HDTV theater, or just putting a TV on the bureau in the bedroom.

✔ **What size do you need?** No, bigger is not always better. You can have a TV that's too large for your space, or too small for your usage. There is an optimal range based on where you intend to place the TV and where you intend to sit.

These first two items — size and budget — will do a lot to narrow your choices before you get to any of the technical or usage criteria, so they are important to nail down first. If you want to fill an 8-foot wall with an image, unless you have a bank account the size of Bill Gates's, you're not going to do that with anything but a front-projection system.

✔ **What do you plan to do with it?** Are you going to be watching a lot of sports events? Movies? Video games? Believe it or not, certain types of HDTVs are better with certain types of content. Sports fanatics will find a big, bright DLP projection system better for their tastes, everything else being equal, while people who watch CNN all day long will want to avoid plasma-screen displays in a big way, due to the burn-in effects of static images (more on this later).

✔ **What will you hook up to it?** If you already have a decent investment in A/V gear, then that gear might dictate certain types (and numbers!) of interfaces or ports on your HDTV system, like these:

- If you have an entertainment system designed around centralized video switching — using a receiver to switch among video sources and destinations — then you're going to need a receiver that can switch HDTV content. That might mean a new receiver, which can be pricey and cut into your budget.

- Do you need a tuner or just an HDTV-ready display — meaning you'll get your HDTV tuner from your cable or satellite company?

✔ **What neat features do you want?** It's easy to be swayed by neat features, but in lots of implementations, you can't access them for various reasons. For instance, if you set up your system so all your signals come in over one cable connection, you might not be able to use your TV's dual-channel features — you could rely on your cable or satellite box for that. (We talk about these issues in Chapter 4.) Still, features are important to all of us, and we'll tell you in this chapter about which ones are the most important.

Budgeting for HDTV

If we all had unlimited funds, we'd simply buy the best of everything. That's why there exists a market for super-high-end gear — those with the big bucks often just buy the top of the line all the time, because they can. (And because they hired consultants to tell them that.)

Most of us are more likely to be working on a budget. Table 2-1, from our *Home Theater For Dummies* book (updated for current rates and an HDTV focus), offers a glance at what you *could* put in your HDTV environment to make it really boom (in a good way).

Table 2-1	Home-Theater Budget Guide	
Role	*Device*	*Price Expectations*
Audio sources	Tape-cassette player*	$100 to $800
	CD player/recorder*	$60 to $600+
	Turntable*	$100 to $5,000+ (really!)
	AM/FM tuner*	$200 to $1,000
Video sources	DVD player	$50 to $1,200+
	VHS/S-VHS VCR*	$50 to $1,000
	D-VHS Digital VCR*	$600 to $1,000
	Digital video recorder (DVR)*	$150 to $1,000
	Satellite system*	$100 to $1,000
	High-definition camcorder	$2,500+
Computer/gaming	Gaming console*	$150 to $200
	Home Theater PC*	$1,800+
A/V system**	All-in-one systems	$200 to $3,000+
	A/V receiver	$100 to $4,000
	Controller/decoder	$800 to $5,000+
	Power amplifier	$500+
	Center, left, right, and surround speakers	$150+
	Rear surround-sound speakers*	$100+
	Subwoofer speakers	$150+

(continued)

Table 2-1 *(continued)*

Role	Device	Price Expectations
Video display***	26- to 40-inch direct-view tube HDTV	$600 to $3,000
	Up to 65+-inch CRT-based rear-projection TV	$1,200 to $5,000
	Up to 65+-inch microprojector rear-projection TV	$2,500 to $6,000
	Up to 120+-inch front projection TV	$3,500+
	32- to 60+-inch plasma-screen TV	$3,000 to $15,000+

Source: Home Theater For Dummies

* Optional

** You don't need all of these parts, just an all-in-one system, an A/V receiver, or a controller/decoder and power amplifier combo

*** Only need one of these displays

Okay, you don't *need* all the gear in Table 2-1. You might be content with just a fabulous HDTV. Or, you might be moving into a new house and want to put in a lot of the above. Your choice.

Table 2-2 gives you a sense of what you might expect to get if you're outfitting an entire room's worth of gear, for different-size budgets.

Table 2-2   HDTV Theater: Bang for the Buck

Price Expectations	Role
$600 to $1,000	This is the "entry level" for HDTV. For this price you should be able to buy a CRT (tube TV) HDTV. You've got a few choices here — you could spend the entire $1,000 and get a relatively big (32-inch) HDTV, or you could go with a smaller one (26-inch or 27-inch) and add in an inexpensive "Home Theater in a Box" system with DVD player and surround sound.

Price Expectations	Role
$1,000 to $2,500	Here you can begin to move up to bigger HDTVs. For a little more than $1,000, you can pick up a nice widescreen 34-inch CRT HDTV; for a bit more, you can start getting into CRT-based rear-projection HDTVs (see Chapter 22). In this price range you can also begin to add fancier home-theater audio solutions, with larger speakers and more amplifier power.
$2,500 to $5,000	This price range is where things start to get fancy! At the bottom of this price range, you can start getting into 42-inch or 50-inch *microprojector* rear projection TVs, like DLP HDTVs. Near the top of the range, you can be considering really big microprojector systems (60-inch or even 70-inch), and you'll also begin to get into front projection systems with HUGE pictures. You can also find a few small (42-inch) plasma TVs in this price range. On the audio side, you can begin to consider high-end receivers and speakers.
$5,000+	Between $5,000 and $10,000, you can build a truly top-of-the line HDTV home theater, with a front projector, the biggest microprojector rear-projection systems, an LCD flat-panel TV, or a big (50-inch or above) plasma TV. Above $10,000, the sky is truly the limit: Imagine high-end "separates" audio equipment, the biggest and best plasma TVs or front projectors, and all the trimmings (including home-theater seating, specialized remote-control systems, even motorized drapes and "movie theater" popcorn machines").

One thing is for sure: Pricing is changing all the time. Two years ago, a lot of the gear mentioned in Tables 2-1 and 2-2 actually cost twice as much as it does currently. As we go to print, Costco (www.costco.com) was selling a 56-inch rear-projection DLP for under $2,400. (Wow!)

Go online to places like www.pricegrabber.com to get a sense of how much the prices have changed since we published this book. Use that benchmark to mentally adjust pricing throughout the book.

In deciding how much to spend overall, we can only give you this advice: Your home entertainment system is probably one of the most-used parts of your home. It helps define your family, social life, business relationships, and so on. These are the places where we personally make a substantial investment because it gets the most use.

Finding the Right Size

We think it's an outright crime that movie theaters can sell tickets for those rows up at the front of the cinema, where your head has to constantly move back and forth (like at a tennis match) to capture all the action. Likewise, the seats far away at the back are just as criminal.

So apply that principle to your home HDTV viewing area. You can be too close to the image (or have the image too large), and you can be too far away (or have the image too small). You definitely know you are too close if you can see the individual pixels on the screen. What you need is "Baby Bear's Just Right" size.

In general, experts determine the optimal size for your HDTV set by dividing the distance you are going to sit from the TV set by 2.5 (don't ask us where they got that number, we haven't a clue — we bet by trial and error — actually there's a lot of science regarding such technically arcane items as the size of pixels and the average person's visual acuity. You really don't want to know!). Table 2-3 gives you our recommendations for common screen sizes, based on your distance from the screen. (Table 2-3 is for widescreen — 16:9 — HDTVs. If you're buying a non-HDTV with the older and narrower 4:3 screen, you can actually sit farther from the screen for the same diagonal size).

Table 2-3	Viewing Distance and Screen Size
Viewing Distance	*Recommended Display Size*
5 feet, 7.5 inches	27 inches
6 feet, 9 inches	30 inches
8 feet	35 inches
9 feet	40 inches
9 feet, 9 inches	42 inches
10 feet	45 inches
10 feet, 5 inches	50 inches
12 feet, 6 inches	55 inches
13 feet, 9 inches	60 inches
15 feet	65 inches

Matching Your HD Needs

You need to match how you intend to use the HDTV with the available technologies. While any HDTV type *can* be used for just about any TV viewing scenario you can think of, certain types of HDTV are better suited for particular uses. This is basically a technical issue — different types of HDTVs use different underlying technologies to create their pictures — which often match up better with some uses than with others.

For instance, if you pack a lot of friends into a wide room to watch movies a lot, you may want to consider a plasma or direct-view CRT (a tube HDTV) display, instead of a CRT rear-projection TV or LCD flat-panel. That's because plasma and direct-view CRTs have the best viewing angle (viewing from the sides). Some other common examples driven by usage are summarized in Table 2-4.

Table 2-4	Finding an HDTV for Your Needs
Use	*Best HDTV*
Home theater	Any projection or flat-panel
Sports	DLP projection or plasma
News and stock ticker	CRT, DLP projection or LCD flat-panel
Portable	LCD or DLP front projector
Gaming	LCD or DLP projection, or LCD flat-panel
Internet/PC	LCD or DLP projection, or LCD flat-panel
Bedroom TV	LCD flat-panel or CRT
Bang for the buck	CRT rear projection
Close to plasma for less cash	DLP or LCD rear projection
Showing off!	Plasma (bigger the better)

Connecting the Other Gizmos

Your HDTV doesn't live in a vacuum. An essential step when choosing an HDTV is to find a model that works within the confines of your home. That means making sure your HDTV works with these:

✓ Your chosen source of HDTV signals

✓ Your existing analog (NTSC) TV signal source

✓ Your existing video-source devices, such as DVD player, VCR, or home theater

✓ The *new* gear you plan to get with your leftover money (yeah right) to supplement your HDTV-powered home theater

A lot of this *is* technical stuff — which we cover briefly in this chapter, and then refer you on to appropriate chapters later in the book with more detailed discussions.

Accessing your HDTV channels

To get the most out of your HDTV, you need to be able to receive HDTV channels. What you need to make this work depends what kind of HDTV you have (or are buying):

✓ **HDTV:** A *true* HDTV contains a built-in ATSC tuner, which can receive over-the-air (OTA) HDTV broadcasts (see Chapter 1 for details on ATSC).

The FCC (Federal Communications Commission) is beginning to require ATSC tuners in all TVs — phasing it in over time, beginning with big-screen (36-inch and bigger) TVs in 2005. So eventually all new TVs will have ATSC tuners — even non-HDTVs, since the ATSC specification includes standard-definition digital TV.

✓ **HDTV-ready TVs:** These TVs can produce HDTV images on-screen, but they don't have an internal ATSC tuner. You'll need some sort of external tuner to pick up HDTV broadcasts.

We're going to use the term *HDTV* as shorthand for both HDTVs and HDTV-ready TVs throughout this chapter — and the entire book. But when you're shopping, keep in mind that not all HDTVs have built-in tuners.

So what do you need? Well, for openers . . .

✓ If you want to watch HDTV from cable or satellite sources, you'll need the appropriate *cable box* or *satellite receiver* connected to your HDTV. (We talk about cable set-top boxes in Chapter 9, and satellite receivers in Chapter 10.)

There are some new *DCR* (digital-cable-ready) HDTVs entering the market that let you watch HDTV over your cable system without a cable box. We talk more about these TVs (and the *CableCARD* that makes them work) in Chapter 9.

✔ If you're going to watch OTA HDTV channels, you'll need two things:

- An HDTV antenna to receive the broadcast signals

- An HDTV tuner (either built into your HDTV, or a separate tuner) to tune into the HDTV channels

We discuss OTA HDTV antennas and tuners in Chapter 8.

Sometimes you may want to mix and match these systems. For example, if you use satellite, you might still use an OTA HDTV antenna to pick up your local HDTV channels.

Unsure whether you want cable, satellite, or local broadcasts for HDTV channels? Chapter 7 helps you make the right choice.

Getting your analog channels

Just because you've bought an HDTV doesn't mean you can *only* watch HDTV stations with it. That would be too frustrating, given that there are still plenty of stations that don't yet broadcast in HDTV. HDTVs are *backward-compatible* with NTSC (the old analog TV system): You can watch the analog channels, and also preserve your investment in NTSC source devices like DVD players, VCRs, and laser discs.

Going native

Most HDTVs have one level of resolution (or occasionally, a couple of them) considered *native* to the TV. That means the HDTV is designed to display images at its specified resolution(s); any signals of a different resolution must be converted (or *scaled*) to the TV's native resolution. (Check out Chapter 1 if you're not sure what we mean by *resolution*.)

For example, many DLP HDTVs have a *native resolution* of 1280 x 720 pixels (720p) — based upon the DLP chip inside these TVs. Your HDTV will probably look its best when it is fed sources that match the native resolution; but it will still look great when the image is scaled to match the native resolution.

More important than native resolution is knowing your HDTV's *supported resolutions* — the resolutions that the TV can actually scale to native. Some HDTVs won't do the scaling that converts between resolutions. This can be important to know when you're connecting an HDTV and an external ATSC tuner (or cable or satellite system). If 1080i is *not* a resolution that your HDTV supports — and that's all your tuner puts out — then you won't get a picture. This is a rare situation, but it can happen.

You can get an HDTV *set* now, and then get the HDTV *channels* later. In fact, because of a device called a *scaler* (discussed in Chapter 5), you might find that your HDTV makes your non-HDTV sources look better than ever. Most HDTVs have a scaler built in, and you can also buy external scalers that are even more powerful.

Most HDTVs — whether or not they contain an ATSC tuner — contain an NTSC analog TV tuner. That way you can plug in an antenna feed and pick up all your local NTSC broadcasts. In the majority of cases, you can also tune in to *analog* cable broadcasts with this NTSC tuner.

Unless you've got a DCR HDTV (discussed in the previous section), you'll need a set-top box to pick up *digital* cable broadcasts.

A few HDTVs — mainly flat-panel plasma and LCD HDTVs — contain no TV tuner at all. Not an ATSC tuner, not an NTSC tuner — nada! If you've got one of these, and you're not using a cable box or satellite receiver, you can use the NTSC tuner built into your VCR to pick up OTA NTSC broadcasts or analog cable.

If you've got an external ATSC tuner for OTA HDTV, it probably also has an NTSC tuner.

Working with your other sources

Chances are very good that you'll be connecting more than just HDTV and analog TV broadcasts to your new HDTV. You'll want to (we're guessing) watch DVDs and videotapes, play video games, and so on.

There are two bits of good news here:

- ✔ All HDTVs will be compatible with the NTSC signals that these devices put out.
- ✔ Most HDTVs will include plenty of inputs on the back (or side, or front) of the HDTV set to accommodate these devices.

Inputs galore! You really don't have to worry about "will my 1982 Betamax work with my 2005 model HDTV?" It will — as long as that beautiful old Betamax itself is still operating (and there's a whole underground world of Betamax enthusiasts to keep you up and running!).

The only real question is how you get all these inputs hooked up and connected to your HDTV. In Chapter 3 we discuss each of the input types themselves in more detail. Beyond understanding the types of inputs, there's a purely *quantitative* angle to this problem.

In effect, you need to count the devices you've got (or anticipate getting) and group them together by the type of inputs they use. Then compare these numbers to the number of inputs on your HDTV. Here's a basic list:

- **Digital inputs:** You're likely to have only one of these (DVI or HDMI) on your HDTV; your HDTV tuner/cable box/satellite receiver or DVD player may use these inputs.

- **Component video inputs:** You probably have a couple of these on your HDTV. Your HDTV tuner, DVD player, and game console (Xbox or PlayStation) can use these.

- **S-video inputs:** You'll probably have a bunch of these (but you'll need them) on your HDTV; your DVD player, VCR, game console, digital cable box, satellite receiver, camcorder, and even PC (yes, PC!) are just a few of the devices that can use this connection.

- **Composite video inputs:** You'll also have a bunch of these; everything we mention here can use this connection method as well.

There's also a *qualitative* angle at play here. In Chapter 3 we explain in detail, but in a nutshell, the connections listed here are shown in order of rank. If you run out of inputs of a certain type (say, component video) and have to use the next one down the list, you lose a bit of video quality. Therefore it can be important, if you've got a lot of gear, to choose an HDTV with *more* digital, component video, or S-video connections, if at all possible.

You can get by with fewer inputs on the HDTV if you use a home theater receiver (see Chapter 19) that provides high-definition *video-switching* functionality. Basically, you can route everything into the back of your receiver, and then use just a couple of cables to connect the receiver to your HDTV. This is a great way deal with running out of the proper kind of inputs on your HDTV.

Which Features Matter?

A bunch of features are listed in the description of every HDTV on the market. Some are important, others are just "bells and whistles" that we don't think make a real difference. (We support diversity, however — if you disagree and think a certain feature is very important, then by all means make that part of your buying decision! This is a very subjective area.)

Here's what we pay attention to:

- ✔ **Picture adjustments:** All HDTVs will give you some degree of control over the picture settings. What we like are HDTVs that let you

  - Set the picture quality differently for each input on the back of the TV — so you can adjust the picture individually for the HDTV tuner, the DVD player, and so on.

  - Save multiple different picture settings in memory (like one for day time and one for night).

- ✔ **Comb filter:** The comb filter is an internal circuit in your TV that separates out the brightness and color information in an NTSC signal before it's displayed on your screen. Look for an HDTV with a *3D* or (even better) a *digital* (also called *3D Y/C*) comb filter.

- ✔ **Front-panel inputs:** Got a camcorder (a MiniDV model, not an HDTV camcorder), or a game console that the kids are always carrying around the house? You'll want some front-panel inputs to connect them to so you don't have to climb behind the TV.

  Look for front-panel inputs that include S-video for better picture quality.

- ✔ **Built-in speakers:** We're *huge* proponents of connecting your HDTV system to a full-up, external surround-sound audio system. We're also fans of low-impact, easy-to-use systems. So, when we want to just watch the news, or turn on that TiVo recording of *Sesame Street* for the kids, we prefer to use the speakers built into our HDTVs. We mention this because some HDTVs (mainly plasmas and LCD flat-panels) don't come with speakers — you'll have to fire up the full surround-sound system for everything you watch.

- ✔ **Surround-sound decoder:** While you need six or more speakers and related amplification systems to get true surround sound (see Chapter 18), you can get improved sound quality for the sound system built into your HDTV if it includes a *simulated* surround-sound decoder, which can create a richer sound from your HDTV's speakers.

Service matters!

When you're buying an HDTV set, you're spending a relatively serious chunk of change. Heck, even if you're loaded, spending $10,000 or more on a front-projector system starts to move out of the pocket-change realm pretty darned fast.

So, it's important to get your money's worth. That doesn't just mean getting the lowest price for the particular TV you've chosen (though that's an admirable goal — one we can definitely get behind); it also means getting decent service to boot.

Now, we've all got our own definitions of what good service is, but here's what we look for:

✓ **Delivery services:** Lots of places (both online and local brick-and-mortar stores) are offering "white glove" delivery services. This means someone delivers that huge big-screen TV that wouldn't fit in the back of your Mini in a million years, and they don't just drop it off on the front stoop and boogie on out of there. Instead they deliver the unit to the room you want it in, get it out of the box, and even take all the packaging materials with them (no doubt, to be humanely recycled!). That's what we want!

✓ Warranty: We've never been huge fans of the "extended warranty" services offered by many consumer-electronics stores. After all, what's the point of paying $30 for a warranty on a $35 cordless phone? But check out the warranty that comes with your HDTV closely — it's a major investment after all. And consider an extended warranty if it's not too expensive. With some types of HDTVs (such as DLP and LCD projection systems) you may pay back the warranty when your bulb needs replacement after a few years of heavy usage.

Chapter 3

Cables and Connections

● ●

In This Chapter

▶ Dealing with digital video cables

▶ Sorting through analog video cables

▶ Coping with copy protection

▶ Plugging into audio

● ●

As you pull your shiny new HDTV out of its (probably very large) box — and consider connecting it to your TV source, your DVD player, your home theater receiver, and all the other "stuff" in your family room or media room — you may have a moment of panic (or at least a small shiver of fear) when you consider how many different choices you have for cables and connectors.

Never fear, *HDTV For Dummies* is here to help you. Cables are actually pretty easy when you think about them in terms of a hierarchy — some cable/connection types are (almost always) simply *better* than others — they give you a better picture or clearer audio. Once you know this hierarchy, you can quickly examine your connection options for any piece of equipment attached to your HDTV, choose your best option, and astound your friends.

In this chapter, we explain analog and digital cable options for audio and video, and explain their positions within this hierarchy. We also explain how *copy-protection* systems may affect your options.

Video Connections

Look at the back of any HDTV (or any DVD player or home-theater receiver) and you see what scares many folks away from jumping in to hook up their own HDTVs and audio/video (A/V) equipment. There are just so darned many choices back there — who could possibly know which connector to use?

Well, *we* know.

High-definition video

There's often a significant difference in the functionality and video quality of connections. Few cables can handle HDTV signals.

Some DVD players have high-definition *connectors*. But when this book was published, *none* of these were HDTV DVD players. Instead, a circuit called a *scaler* (see Chapter 6) converts the *standard* DVD picture to work a little better on an HDTV screen. It's a worthwhile improvement, but you probably should use all your high-definition connections for *true* HDTV components before you use a high-definition connection on a DVD player.

Digital connections

Digital video connections (such as DVI-D, HDMI and FireWire) are the best choice for often-used HDTV video connections, such as the link to your HDTV from a satellite or cable receiver.

Not all HDTV devices use the same digital connections, but it's usually worth the trouble to use digital connections when you can, even if you need an adapter for *different* digital connections. As the HDMI connection replaces DVI-D, you may find yourself in a situation where you need to use an inexpensive (under $50) adapter. For example, if your HDTV has an HDMI connector and your HDTV set top box or tuner uses DVI-D, you can connect your devices through one of these converters.

The first rule of HDTV connections: Use a *digital* video connection to connect from a source device to your HDTV, if you can. Digital connections almost always provide the best picture.

Theoretically, HDMI connections can offer the best picture quality of any of the digital connections available on HDTVs, simply because these cables have so much *bandwidth* that they can offer *uncompressed* HDTV signal transmission. In the real world, however, any of these digital connections offers an exceptionally clear and sharp picture, and HDTV signals are almost always compressed for transmission or storage anyway.

On some pieces of gear (typically HDTV cable set top boxes — discussed in Chapter 9), some of the digital connections may be disabled — the connections are physically present, but the software within the device that lets them work is turned off. So, for example, if an HDTV set-top box from your cable company has a FireWire port, you probably can't use that FireWire port to connect a D-VHS recorder. Check with your cable company before spending money on cables and equipment that use these ports!

DVI-D

The most common digital-video cable for HDTV is *DVI-D.* (DVI stands for *digital video interconnect,* and the extra D means it's for digital TV.) Figure 3-1 shows the DVI-D connector for HDTV.

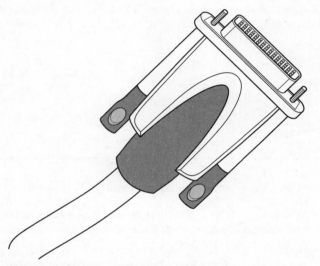

Figure 3-1: DVI-D your way to high-def video.

If you're using DVI-D connections, watch out for two problems:

✔ Not all DVI connectors work with HDTV. Make sure you have *DVI-D* cables if you use DVI-D in your HDTV system.

Computers use another type of DVI connector that has a confusingly similar name: *DVD-I.* The DVD-I connector has five extra pins (four pins around a central crosshair-shaped pin) on one side; these send analog video signals from computer video cards to computer monitors. You typically won't find DVI-I in home HDTV systems. (There are some projection systems that are also used with computers to beam PowerPoint slides up onto the conference room wall.) You can use an inexpensive adapter to connect a DVD-I cable to the DVD-D receptacle on your HDTV — but you'll only be able to receive digital video signals that way, not analog.

✔ DVI-D connections often require the HDCP copy protection system for true HDTV video performance. If just *one* of your HDTV components doesn't have HDCP, you may not get true HDTV performance from DVI-D connections. (HDCP is explained in the sidebar, "No copying!")

DVI-D is the only digital HDTV connection that *can't* carry *audio*.

HDMI

The *HDMI* system (for *High-Definition Multimedia Interface*) is specially designed for HDTV connections. It carries both *HDTV video* and *digital surround sound*. Figure 3-2 shows the business end of an HDMI cable (the connector, in other words).

Figure 3-2: HDMI: the latest and greatest.

HDMI has a couple of advantages in an HDTV system:

- ✔ You need only *one* HDMI cable to connect both HDTV video and surround-sound digital audio signals.

- ✔ HDMI is an extremely high-bandwidth technology (5 gigabits per second). It has extra bandwidth to accommodate *future* HDTV formats.

HDMI connections often require the HDCP copy-protection system for true HDTV video performance. If you're using any HDMI connections, make sure that *all* your HDTV components support HDCP. HDCP is explained in the sidebar, "No copying!"

FireWire

FireWire is the least-used HDTV connector. It can transmit both video and audio. Figure 3-3 shows a FireWire connector.

FireWire is the only *two-way* connection for HDTV — the same cable can send HDTV video (and audio) to *and* from devices. This two-way connection is great for HDTV recording systems — for example, one cable fully connects an HDTV with a D-VHS VCR.

We use the name FireWire throughout this book, but the same system is known by a couple of other names:

- ✔ Engineers and nerds call FireWire the *IEEE 1394 Standard.* (IEEE is the *Institute of Electrical and Electronic Engineers.*)

✔ Some manufacturers use the name *i.LINK* instead of FireWire.

Consumer-electronics manufacturers usually prefer a snappy name like *i.LINK* or *FireWire* over something boring like *1394* (not much you can do to liven up that number, huh?).

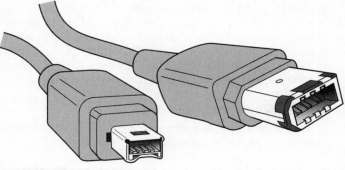

Figure 3-3: The Wire's on Fire!

No copying!

Film and television studios are worried about people copying their programs and distributing them to others. HDTV (and DTV in general) is worrisome to the studios because there may be "perfect" digital copies. (Non-digital copies get worse when they're recopied, so they aren't such a threat.)

These content owners have lobbied the government and manufacturers to include *copy-protection* systems in HDTV devices, such as tuners, set-top boxes, satellite receivers, and HDTVs themselves. These systems can keep you from making copies (or sometimes even *one* copy) of any HDTV program.

The most common copy-protection system is *HDCP* (high-bandwidth digital content protection) — this system is in new HDMI-equipped devices and many DVI-equipped devices. HDCP *encrypts* (scrambles) the content sent between devices like tuners and TVs. This encryption is a problem if you have a DVI connection where one piece (like a set top box) uses HDCP, and another (like your HDTV) doesn't. That's because the content owners have rigged the system to *downres* (down-resolve) HDTV programs to standard-definition unless HDCP is present at all points in the system. You could end up unable to get an HDTV signal on your HDTV!

If *some* but not *all* of your HDTV devices have HDCP, then component video may be the only way to get a true HDTV signal between your devices. Component video isn't limited by HDCP, because component video isn't a *digital* signal.

We're big fans of FireWire, but most HDTVs and HDTV devices use either DVI-D or HDMI connections instead of FireWire. The most common place to find FireWire is a D-VHS VCR.

FireWire isn't part of the HDCP copy-protection system. (HDCP is explained in the sidebar, "No copying!"). Instead, FireWire uses *its own* copy-protection scheme called "5C-DTCP" (or *5-company digital-television content protection*), which provides similar protection of content that the big TV companies don't want you to record for yourself. The 5C system (that's the industry shorthand for it) basically acts just like HDCP, letting only authorized (5C-equipped) equipment make recordings of "flagged" material.

Analog component video

Component video is the only *analog* video-cable connection that can handle HDTV or progressive-scan DVD signals.

S-video and composite video don't carry progressive scan.

Technology

Component video is a set of *three* analog cables, as shown in Figure 3-4.

The component video signal is divided into three (you guessed it!) *components:*

- ✔ *Y* is the *luminance* (brightness) signal.

- ✔ *Pb* and *Pr* each carry part of the picture's *chrominance* (color) information. Your TV uses these two chrominance signals to create the red, green and blue colors that can be mixed together to create any color on your display. (Sometimes, Pb is labeled *B-Y* and Pr is labeled *R-Y.)*

Connection

Component video connections use three normal analog cables.

If you already have three standard *composite video* cables, you can use those cables instead of a set of "official" component video cables. There's no functional difference between the two, though many folks find it convenient to buy component video cables bundled together (so all the cables are neatly labeled, attached in the right order, and don't get lost).

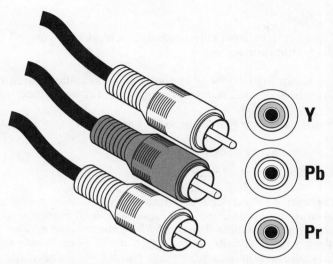

Figure 3-4: The best analog video connection — component video.

If you're routing all of your video cables through a *home theater receiver* (as covered in Chapter 19), check your receiver's specs before routing component video connections through it. The receiver's *component video bandwidth* specification should be

✔ At least 10 MHz for *progressive-scan DVD players*

✔ At least 30 MHz for *HDTV connections*

 Component video may be the *only* connection that allows a true HDTV signal in your system if *some*, but not *all*, of your HDTV components use the HDCP copy-protection system. (HDCP is explained in the sidebar, "No copying!")

Traditional video

You're probably not connecting only HDTV sources into your spanking-new HDTV. Most of the today's TV content is *standard* definition in both *broadcast media* (over-the-air, cable, and satellite) and *prerecorded media* (DVD and VCR).

Most source devices for standard-definition video use one of the following traditional *analog* connectors, not the high-definition video connections mentioned earlier in this chapter.

S-Video

The best traditional video connection is *S-Video*. Figure 3-5 shows the S-Video connector.

If you can use an S-video connection, you should (unless you can use a high-definition *digital* or *component video* connection).

S-video cables transmit the video signal in two separate channels:

- ✔ Luminance (brightness data)
- ✔ Chrominance (color data)

Separating luminance and chrominance can deliver a better picture because it bypasses a circuit found within all but the smallest and cheapest televisions (the *comb filter*). Because an S-Video cable already uses separate conductors, your TV doesn't have to separate this information out with its own comb filter. Usually this built-in separation provides a better picture on the TV.

S-video connections often are found on DVD players, game consoles (such as Xbox), and many satellite and digital cable receivers or set-top boxes. Some VCRs also have S-video connections.

Figure 3-5: Use S-Video for non-HDTV or progressive-scan sources.

S-video cables are a bit tricky to connect:

✔ An S-video plug has four *very, very* delicate pins and a small plastic tab. It's incredibly easy to misalign these pins and bend them (and even break a pin off). Easy does it. (But you knew that, right?)

✔ A properly aligned S-video plug easily slides in. The key is to line up the small plastic tab with the corresponding slot on the jack, and *gently* push. *Don't twist!* If you're pushing hard, it's probably not aligned straight (and you're probably bending pins). If the plug is hard to push in, stop and realign it.

Composite video

The oldest basic connection for separate video units is *composite video*. Composite video cable is a 75-Ohm cable with plugs called *RCA connectors* (the same connectors used by most audio cables).

Composite video cables (shown in Figure 3-6) carry the entire video signal on a single conductor. They don't carry any audio — you need separate analog or digital audio cables for that.

Composite video cables can offer a much better picture than a standard coaxial "antenna" connection. However, composite video really isn't the best choice for higher-resolution analog sources, such as *S-VHS-C* and *Hi8 camcorders, DVD players,* and *videogame consoles.* Use an S-Video connection when you can.

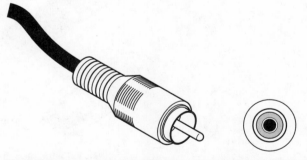

Figure 3-6: Use composite video for low-resolution video sources.

Coaxial cable

Coaxial video cable (*coax* — pronounced *CO-ax* — for short) is the cable that's probably either running down from your attic antenna or coming into the side of your house from the cable company or satellite dish. It carries both video and audio signals.

A video system should only use a coax connection when you are connecting something directly to an *outside* feed (an antenna, satellite dish, or incoming cable line). Whenever you connect a device to your HDTV, any other video connection (even S-video or composite video) should give you a better picture than coax video.

The connectors on the ends of coax are *F Connectors,* as shown in Figure 3-7. An F connector has a small *pin* (the *conductor)* sticking out of the middle, and a metal *barrel* around the outside.

Use *screw-on* F connectors if you can. Screw-on F connectors can make a much better (tighter) connection than push-on connectors.

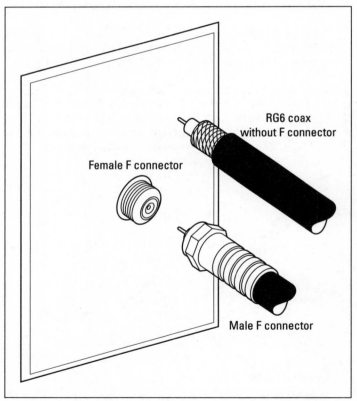

RG6 coax
without F connector

Female F connector

Male F connector

Figure 3-7: Use F-connectors to hook up your cable or antenna feeds.

Audio Connections

Audio connections aren't as complicated as video. There are basically three choices, and only two deliver digital surround sound.

Every video connection for your HDTV needs *separate* audio connections except *coaxial cable* (which you should only use for *incoming* feeds from antennas, cable services, and satellite dishes), *HDMI,* and *FireWire.*

Digitizing your audio

Digital audio signals (such as those put out by DVD players and HDTV tuners) are a digital stream of bits from the source device to either your *receiver* or *HDTV,* which decode this bitstream.

If you have digital connections available between two pieces of gear, use *digital,* not *analog:*

 ✔ Some home-theater gear (like DVD players or HDTV tuners) *requires* digital audio for the most advanced surround-sound systems (for example, Dolby Digital). If you use analog, you'll revert to a less effective surround-sound standard.

 ✔ The *longer* digital audio stays *digital,* the better.

The biggest advantages of digital audio are

 ✔ Near *immunity* to interference

 ✔ A *"pure"* digital signal all the way to the receiver

Digital audio can use either digital coaxial electrical cables or optical (*Toslink*) fiber-optic cables.

There's very little difference in performance between optical and coaxial digital cables. The decision usually revolves around which system both your *source devices* and *receiver* happen to use.

Digital coaxial

Digital coaxial cables look basically identical to analog audio cables. They're made of similar materials, and they use the familiar RCA-plug connector. The internal construction is different, however — and coaxial digital-audio cables are designed to provide 75-Ohm impedance (just like video cables).

We've tried, and can't hear the difference, but thought we'd let you know that *some* HDTV enthusiasts think coaxial cables *sound* best.

Optical (Toslink)

Toslink optical cables are made of fiber-optic cabling (usually plastic fiber). They transmit the digital audio as pulses of laser-generated light, not as electrical signals.

Toslink cables are particularly immune to *electromagnetic interference*.

The connector on a Toslink cable is quite distinctive, as shown in Figure 3-8. It looks like nothing else in the world more than the profile of a house, with a small pin (actually the end of the fiber) sticking out of the side.

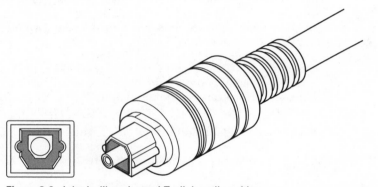

Figure 3-8: It looks like a house! Toslink audio cables.

The "female" Toslink connector on your equipment may have a *dust cap* to keep the optical pickups clean while it isn't connected. Remove (and *save!*) this dust cap before you try to connect.

Attaching analog audio

A couple of familiar audio connections still do a lot of the work in today's entertainment systems.

RCA cables

The basic building block of any audio connection is the tried-and-true analog audio cable (often called the *RCA cable*), as shown in Figure 3-9. You'll find analog audio cables on the back of almost every source device you connect to your HDTV — ranging from HDTV tuners to 25-year-old VCRs.

Generally, you should use analog audio connections only if you can't make a *digital* audio connection.

Analog audio cables are used

> ✔ In pairs (for basic stereo sources)
>
> ✔ Alone (for connecting a subwoofer to a receiver)
>
> ✔ In sixes (for connecting either a DVD-A or SACD player to the receiver, or connecting the external surround-sound decoder on a DVD player)

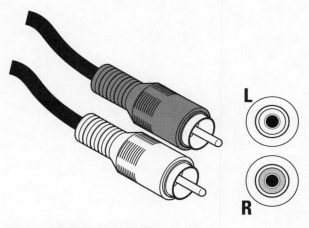

Figure 3-9: Analog audio cables.

Headphone and microphone jacks

Some video sources (such as computers and camcorders) have *headphone jacks* and *microphone jacks* instead of RCA-type audio connections.

If your gadget didn't come with adapters to connect from these jacks to an audio component that uses RCA-type connections, the nearest electronics store usually has the adapter you need.

Part II
Love at First Sight

The 5th Wave   By Rich Tennant

WARREN ENJOYED SHOWING OFF HIS NEW 52" FLAT PANEL HDTV DINNER

Dot's Veal Cutlet
Peas and Yams

In this part . . .

*W*ell, if this were a book about books, we'd say you can't judge a book by its cover — but it's a book about HDTV, so we have to revise this to say, "You can't judge an HDTV by its display." There's so much about an HDTV system that is more than skin deep, and this part of the book helps you understand how to optimally install, calibrate, and scale your HDTV system.

We start out with a chapter devoted to using all the connections we introduce in Chapter 3. We help you figure out how to connect your HDTV to over-the-air, cable, and satellite signals, as well as how to bridge in your DVD, VCR, DVR, camcorder and other sources of video content. We also tell you how to send signals out to your home surround sound stereo system, so you can truly enjoy your HDTV theater.

Then we discuss ways to optimize this system, by making sure your signal is properly calibrated for your specific home environment. As we just mentioned, you can't judge an HDTV that's fresh out of the box by its display, because most of them are improperly calibrated (the picture's not optimally adjusted, in other words).

Then we wrap up the part with a discussion on how to optimize the signal coming into your HDTV, using video processors and scalers. Aside from being fast ways to clean fish, scalers help you match your input-source signals to your best HDTV display format. Without a scaler (either internal or external to your display), the images coming in from your video sources might not be properly aligned with the best display characteristics of your HDTV.

By the time you are done with this part, you can rest assured that your HDTV is just the way it should be. Isn't that a comforting feeling!

Chapter 4

Hooking Up Your HDTV

*S*etting up your HDTV does not have to be a complicated process. In fact, you should be able to test and use your HDTV with minimal setup if you wish — it's only when you start adding DVD players and home-theater systems that the connections get a little spaghetti-like.

In this chapter, we help you open up the box and get the HDTV out (But of course you really don't believe that. Oh well. Call Pat — he'll come right on over! Kidding.). Okay, so what we *really* help you do is understand the overall layout of your HDTV system, and to understand where the parts fit. We start small and simple, with just an over the air connection, and work up to your most intense and complex HDTV home-theater operation.

Throughout this chapter, we refer to all the connections discussed in Chapter 3. You can flip there for the pros and cons of a particular cable or interface.

The exact ports, connections, options, and other attributes of your HDTV set will vary by manufacturer, model, country, and so on. So we're going to use a rather generic diagram to illustrate your HDTV set's interfaces, as shown in Figure 4-1. In practice, you may not have all of these, or they might be scattered around on the sides of your HDTV (as often happens with a plasma) or across the back (as you'd get with a projector). The key point is what you are connecting — and with which cables — so don't sweat it if Figure 4-1 doesn't look exactly like your TV.

Making Connections

Unless your HDTV has a built-in DVD player or VCR, you need to make some connections to actually use it — the video signal must come from somewhere. There's probably a cluster of input and output jacks (*jack panel*) on your HDTV. How you use these jacks is determined by the devices you connect to your HDTV.

Jack panels

Depending on the cabinet design, there may be jacks on the front, at the side, or on the back of your set. Figure 4-1 shows the input and output jacks you're likely to find on the back of a CRT or projection HDTV.

DVI-D, composite-video, S-Video, and component-video connections carry video information only. You need separate audio cables to carry the corresponding audio signals. Chapter 3 covers these connections.

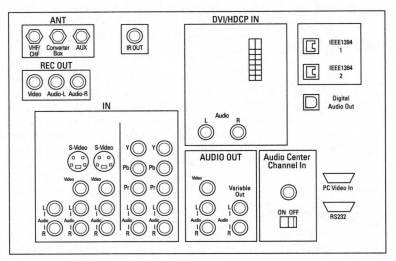

Figure 4-1: Your master set of HDTV ports.

Input jacks

Input jacks only *receive* broadcast signals and programs from your audio/video devices so you can watch them on your HDTV. You'll probably see most or all of the following audio/video input jacks on your HDTV, as shown in Figure 4-1:

✔ **ANT-IN:** Two or more ports for

- NTSC analog and ATSC off-air signals

  These ports only work for ATSC if you've got an HDTV with built-in HDTV (or ATSC) tuner.

- Analog and digital cable-TV signals

  These ports only work for digital cable if you've got a DCR (digital-cable-ready) HDTV (see Chapter 8).

✔ **DVI-D/HDCP IN:** A digital-video input, usually teamed with two R/L inputs for audio

DVI-D/HDCP ports can't plug into cables that are connected to a PC's (similar) DVI-I connection.

✔ **Video In:** Usually these are in sets with

- Composite and S-Video video inputs

- Standard audio inputs

Usually these are connected to composite or S-Video-equipped video systems like VCRs.

Your HDTV may require you to tell it whether you're connecting to the composite video or S-Video jack. (Check your owner's manual.)

✔ **Component Video In:** Component video plus standard audio inputs for accepting signals from component video systems such as DVD players

✔ **PC inputs:** Usually divided into

- **PC Audio Input:** These audio jacks connect to the audio output ports on your PC.

- **PC Video Input:** These video jacks connect to the video output port on your PC.

Output jacks

Output jacks *send* signals from your HDTV to your audio/video devices so you can

✔ Record, listen to, or distribute the programs

✔ Control other devices

You'll probably see most or all of the following audio/video output jacks on your HDTV, as shown in Figure 4-1:

✔ **IR Out:** An Infrared port for sending IR signals to control your attached devices

✔ **Audio/video outputs:** Usually an HDTV has two kinds of audio/video outputs that do a range of tasks:

- **REC Out:** A "record out" connection for recording what you see on your HDTV to an analog VCR.

- **A/V Out:** Regular composite video and standard audio outputs for connecting such devices as a VCR for editing and dubbing. Usually this output is *bridged* directly to an input; whatever is connected to the input jack goes to this output.

These audio/video outputs usually have a couple of limitations:

- They output a *downconverted* video signal, not HDTV.

- You can't adjust the audio volume with the TV remote.

✔ **Audio Only:** HDTVs usually have a couple of outputs for sending audio to other devices, such as amplifiers, receivers, and decoders:

- **Digital Audio Out:** A digital audio connection (usually an optical "Toslink" connector) for connecting external Dolby Digital-enabled amplifiers, receivers, decoders, or other home-theater systems that receive optical audio.

- **Variable Audio Out:** These are standard audio ports for connecting an analog amplifier with external speakers.

Variable audio allows you to adjust the volume of your external sound system with your TV remote.

Many TVs have an on/off switch that governs how the onboard speakers are used. You may be able to switch your speakers so either

✔ The internal speakers carry all the normal audio signals.

✔ The TV's audio goes directly to the A/V receiver, and either

- The TV's speakers can be the A/V receiver's center speakers for surround sound.

- The TV's speakers are entirely off.

Bi-directional jacks

Bi-directional jacks both send and receive data. A couple of these jacks are common on HDTVs, as shown in Figure 4-1. Two types exist:

✔ **FireWire/IEEE 1394:** These ports are for connecting devices for compressed video and audio signals.

HDTVs with IEEE-1394 FireWire ports don't always operate with all the other devices that include such connections. For instance, usually you can't hook up a 1394-outfitted Mini DV camcorder to a TV. Check your manual closely about what can be connected to this port.

✔ **RS-232 Jack:** This is a serial-connection port that can be used with your PC for data transfer (like *firmware* upgrades), and which can also be used with some automation systems.

Connecting basic TV sources

When you get your HDTV home from the store, we know you'll want to turn it on and fire it up. You don't have to connect *everything* right from the start. The following connections are the most basic connections you can make with your HDTV.

Throughout this chapter, we deal with VCRs and standalone DVRs as interchangeable units — they connect to your HDTV in the same fashion. If you have both, see the section called "Two VCRs for editing."

If you want to connect your VCR or DVR right now, skip to "DVR and VCR connections" later in this chapter.

Dealing with the "No Picture" Scenario

Most HDTVs have a fairly large number on inputs on the back (and usually some on the front too!) that can accept connections from antennas, cable TV feeds, cable set-top boxes, satellite receivers, DVD players, DVRs and more. Having all of these connectors is a good thing, but it can occasionally cause a bit of trouble.

So once you connect one of your source devices to your HDTV, you may run into a situation where you start up your HDTV and expect a picture, and get . . . well . . . nothing. Don't panic. Most TVs won't automatically search amongst all the inputs you've got hooked up to find the one that's "active". So read the manual for your shiny new HDTV, and as a first step — before you learn anything fun (such as how to operate the picture-in-picture function or the "zoom" function), figure out how to select the various inputs.

Some HDTVs have *assignable inputs* so you can tell your HDTV and remote control that, for example, Input "1" on the back (which may be a component video connection) belongs to the DVD player and should be activated when you press the DVD button on your remote, and so on. It takes only a few minutes, and it's worth getting figured out right away.

Antenna or unscrambled cable

The simplest connection you can make is connecting an antenna into your HDTV — with no worries about cable boxes, VCRs, DVDs, home theater receivers or anything else. You use a direct antenna connection when

- ✔ You don't need an external TV tuner to tune into NTSC or HDTV channels
- ✔ You don't need a cable box or satellite receiver to unscramble channels
- ✔ You don't connect a VCR or DVR

If you have cable TV and all your channels are *unscrambled* — meaning you don't need a cable "converter" or set-top box — you can connect using this same method.

This connection is shown in Figure 4-2.

When you use an antenna to pick up HDTV or standard-definition broadcasts, it is called using an *OTA* (or over-the-air) connection.

If you are using an indoor antenna to grab either HDTV or regular, standard-definition OTA signals, make sure you keep the antenna away from the TV to avoid noise on the screen.

Cable box

Your best cable-television connection depends on how (or whether) you need to use a cable box to access cable channels.

If you don't need a cable box, follow the instructions in the preceding section.

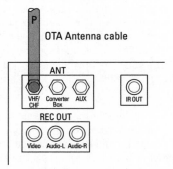

Figure 4-2: The most direct HDTV connection.

Cable set-top box

This option (Figure 4-3) adds a cable-set-top box (or cable set-top box) to the equation. You would use this connection plan if

- ✔ Your cable company requires a cable set-top box only for scrambled or digital channels.

- ✔ You don't connect a VCR or DVR.

Even though most cable set-top boxes have an "RF" output that can connect to the ANT (antenna) input on the back of your HDTV, you'll be better off using the composite video or S-Video connections on your cable box and HDTV, if they're available. (See Chapter 3 for more details on these connections.)

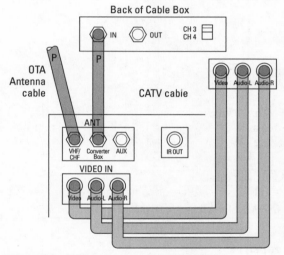

Figure 4-3: Cable set-tops and your HDTV.

The HDTV's internal converter eliminates the need for an external splitter so you can switch between two options:

- ✔ Unscrambled signals come straight into the TV set.

- ✔ Scrambled signals come in through the cable set-top box.

Cable set-top box

This option (Figure 4-4) consolidates all your signals over one link to the HDTV set. You would use this connection option if

- ✔ Your cable company scrambles all its channels, requiring you to have a cable set-top box.

This situation limits how much you can actually use TV-set features that might allow you to view two channels at once (that is, *picture-in-picture* capability). If the cable-set-top box sends one channel at a time, that's all you can watch.

✔ You don't connect a VCR or DVR.

If your HDTV is DCR (Digital Cable-Ready), you may be able to use a *CableCARD* and skip the cable box. We discuss this choice in Chapter 8.

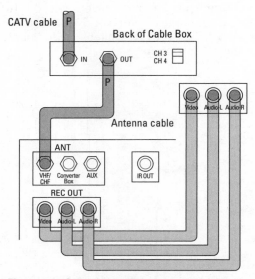

Figure 4-4: Going cable all the way, all the time.

Satellite receiver

This option (Figure 4-5) can consolidate all your signals over one link to the HDTV set if you like. You would use this connection option if

✔ You have a satellite service and receiver.

The signal from satellite dish connects to the receiver, which in turn connects to the HDTV.

✔ You have an over-the-air antenna.

The antenna cable connects to either

• The HDTV (via one of its antenna ports)

• The satellite receiver

✔ You don't connect a VCR or DVR.

Most new satellite receivers have component-video connections, in addition to composite and S-Video connections. Component is always the best of these three options for this application.

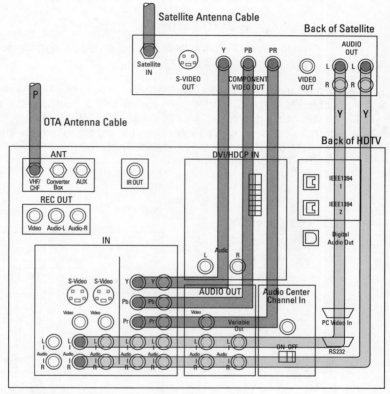

Figure 4-5: Beam me down, HDTV-style.

Connecting DVRs and VCRs

If you have a DVR or VCR, then your connection options increase. The upcoming subsections describe some potential connection situations you might encounter.

In this section we cover the DVR and VCR as interchangeable items — which is exactly what they are, functionally speaking.

If your cable or satellite set-top boxes or receivers has a built-in DVRs — you don't need to make any extra connections at all. All the connections are internal to the set-top box or the receiver.

VCR with antenna or unscrambled cable

The most basic VCR option (Figure 4-6) is to run an antenna feed or cable TV connection to the VCR, and then on to the HDTV. You would use this connection if either of the following is true:

✔ You're using an OTA antenna.

✔ Your cable company does not scramble its signals (so you don't need a set-top box).

If you have both an off-the-air antenna and cable TV, connect the cable TV to the UHF/VHF (also called ANT 1) jack and the antenna to the AUX (also called ANT 2) connection. While this might seem counterintuitive, most TV features are set up to use the UHF/VHF connection as the primary default. That's why you'd want your cable connection on this port.

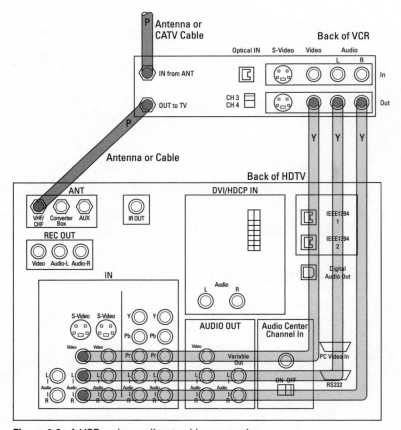

Figure 4-6: A VCR and your direct cable connection.

VCR with a cable-set-top box

This VCR option (Figure 4-7) is to run a cable TV connection to the VCR, and then on to the HDTV. You would use this connection if

✔ Your cable company scrambles some of its signals but not all of them

✔ You connect a VCR

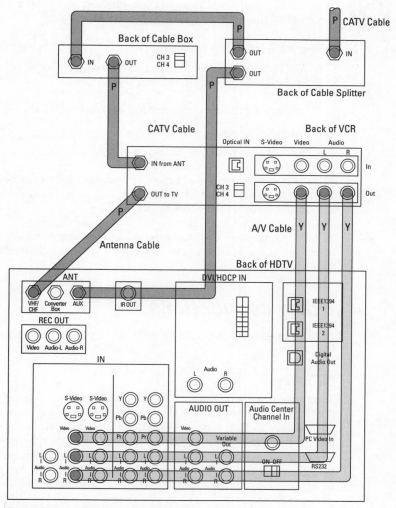

Figure 4-7: A VCR and cable set-top in tandem.

A splitter takes the signal from the cable company and sends identical signals on two paths:

- ✔ Directly to the TV's ANT-2/AUX antenna port
- ✔ Through the cable set-top box to the TV's S-Video or composite video inputs

 Using two antenna connections lets you use the TV's tuner on unscrambled channels for such features as picture-in-picture. The cable set-top usually only sends one channel at a time, so you can't use most of the TV's multichannel features with it.

VCR with a satellite receiver

This VCR option (Figure 4-8) is to run a satellite TV connection to the VCR, and then on to the HDTV. You would use this connection if

- ✔ You use a satellite company for your main signals
- ✔ You have a VCR
- ✔ You have an over-the-air antenna

Most new satellite receivers have component-video connections, in addition to composite and S-Video connections. Component is always the best of these three options.

In this arrangement, the VCR can be used to view or record antenna-based channels, in addition to viewing the antenna channels through the HDTV.

Other connections

The fun does not have to stop here. You can add more to your HDTV setup.

DVD player

You can add a DVD player to the cable TV service, antenna, satellite, DVR, VCR or whatever it is you have hooked up to *your* HDTV quite easily. In fact, it doesn't really matter what you've got connected to your HDTV — the DVD player (and other source devices like videogame consoles or laserdisc players) will connect to their own set of audio and video inputs on your HDTV. The key think to remember is to remember to use the highest quality video and audio connection available to you (see Chapter 3 for details on this). Whenever possible, use component video and digital audio cables — move down from there based on what's available on your HDTV and DVD player.

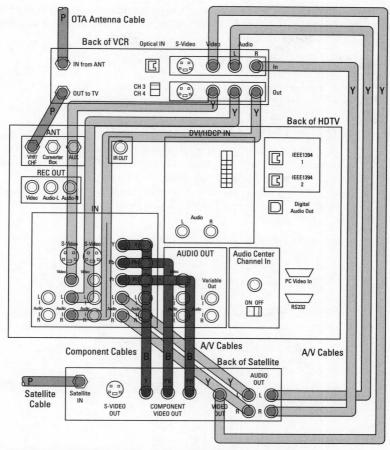

Figure 4-8: A VCR mated to your satellite receiver and HDTV.

If you have a VCR/DVD combination unit, you probably have to make separate connections for the DVD player and the VCR, just as if they were separate devices.

Two VCRs for editing

If you do a lot of taping and editing, then you might have two VCRs connected together (as in Figure 4-9). You can either

✔ Connect two VCRs *serially*:

- The playing VCR's Line Out (video and audio) ports connect to the Line In ports on the recording VCR.

- The recording VCR's Line Out ports send the signal to the HDTV's Video and Audio In ports.

✔ Hook up the playing VCR to the Line In ports on the HDTV, and then connect the recording VCR to the Line Out jacks of the TV.

On most TVs, the Video Out jack does not output a picture-in-picture frame, so you can't record that image.

If you're connecting two VCRs together for tape editing, you still cannot edit a tape that has copy protection.

Digital cable or satellite connections

If your set-top box is digital, you most likely have component video connections — and potentially even DVI/HDCP connections (check out Figure 4-10). In either case, you have to connect audio cables separately. The rest of the connections to other devices are the same as discussed elsewhere in this chapter.

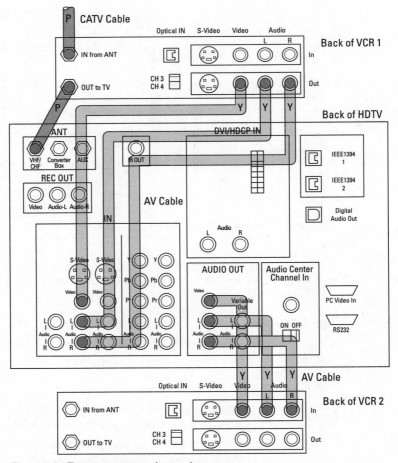

Figure 4-9: Two, two, two are better than one.

Getting your PC hooked up

If you have a Media Center PC (or a homemade HTPC — home-theater PC), you'll definitely want to connect it to your HDTV. How you make this connection depends upon two things:

✔ What kind of video card you have in your PC

✔ What kind of connections are available on your HDTV

The video card determines what kinds of connections are available on the PC; the back of your HDTV determines the other end of the connection. Some HDTVs can connect directly to PCs, and have a *VGA* connection. If you don't have one of these connections, you'll want to use component video or S-Video cables to connect your video, as well as a pair of analog audio cables.

If you have either a Media Center PC or a PC with DVR software, you can treat your PC like a VCR or DVR (as described in this chapter).

 To make sure your DVI/HDCP device resets correctly, we recommend that you turn on the TV first, then power-up the DVI/HDCP gear — and when you turn them off, shut down the DVI/HDCP device first.

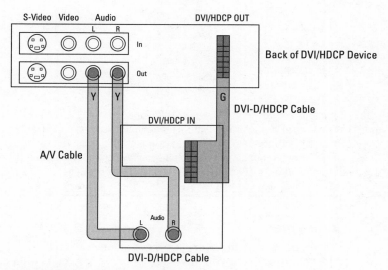

Figure 4-10: Going with a digital set-top box.

Picture-in-picture with cable boxes

When a cable system requires you to use a special cable box (for example, to access premium, digital, or pay-per-view channels), your TV receives only one channel at a time from the box. That means you can't use some of your TV's special features — in particular, picture-in-picture. If all your cable channels (or satellite channels) require the box, then there isn't much you can do about it. (You may be able to get a cable or satellite box with these features — a *two tuner* model — if you want them.) But if your cable system has some *unscrambled* channels, you can add an extra cable input directly to your TV that lets you use all of your TV's features on unscrambled channels. (You can set up your system this way with any of the cable-box connections shown in this chapter.)

This arrangement requires a cable splitter and a couple of extra 75-ohm cables. Instead of connecting the incoming cable feed directly to the cable box, you connect the incoming feed to the splitter. The extra cables connect the TV and cable box inputs to the splitter. The following figure shows this extra connection with a cable box. Note that with this connection, the TV has direct access to any unscrambled cable channels for such features as picture-in-picture.

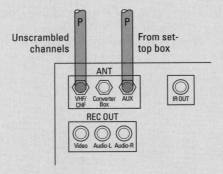

Basic channels are hooked up to an Antenna input on your TV. To tune these channels in using your TV's picture-in-picture functionality, just select that antenna input. You can watch *all* your channels — basic, unscrambled channels, and premium channels — through your cable box, which should be connected to your HDTV using composite or S-Video cables (along with a pair of audio cables).

Camcorder

These days it seems everyone's got a camcorder. If you're part of the "in crowd" and you have a camcorder, you can connect to your HDTV via S-Video if you have it, composite connector if you don't. (Figure 4-11 shows what this setup looks like.)

The audio outputs of your camcorder will connect to a pair of analog-audio inputs on your HDTV.

Most HDTVs have *front-panel* inputs (usually one for S-Video, one for composite video, and a pair of analog-audio inputs) designed especially for connecting camcorders — you don't have to bend over in an unflattering way behind the TV to make your connection.

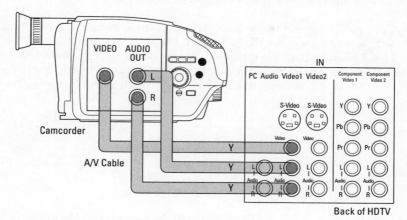

Figure 4-11: Watching your home movies on the HDTV.

Surround-sound system

You can — and we hope you do — connect your HDTV to a full-fledged home stereo system. This is shown in Figure 4-12. You would use this scenario if

✔ You want full surround sound from your HDTV experience

✔ You have a home-theater A/V system rated for surround sound

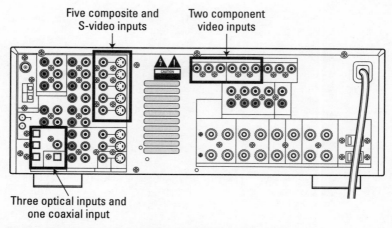

Figure 4-12: Going with the home theater hookup.

This is actually quite simple in design — you're taking the video and audio out from the TV and using them to drive what comes out of the stereo system. Note that if you're using your receiver as a video switch, then you would instead route the video signals into the receiver *before* they go on to the HDTV.

If you plan to use a receiver as a video switch to control the signals going to your HDTV, know that the cables used to carry HDTV, DVI and HDMI, are typically *not* supported by receiver video switching, so you'll need to make those connections directly from source to HDTV, and not use the receiver's video switching. More on this in Chapter 19.

Chapter 5

Enhancing Your HDTV

*W*hen you bring your HDTV home, take it out of the box and turn it on, it will be in what many experts call *torch mode:* the brightness of your picture will be cranked way, way up. Manufacturers do this because of the retail environment — lots of bright overhead lights that could "wash out" the picture on the showroom floor. Manufacturers also crank up the picture brightness because, when ten or twelve HDTVs are lined up next to each other showing the same video, the brightest picture tends to catch your eye (so you think, "Must buy *that* one!").

In this chapter, we tell you about the different adjustments you can make to your HDTV's picture. We also tell you about a series of different inexpensive DVDs (and a pair of D-VHS tapes) you can buy to really get things set up correctly — just adjusting the picture by eye-balling it isn't going to cut it. We even tell you about some free DVD software that can do most (but not all) of what the other DVDs can do. Finally, we tell you about the ultimate in HDTV maintenance — like a spa day for your TV — the professional calibration.

Why You Need to Calibrate

Unfortunately, the "torch" mode your set was adjusted to when it came out of the box *can't* give you the best picture quality when that HDTV is in your home. In fact, it gives you a *lousy* picture in a darkly lit room (which is best for movie-watching). Even in a normally daytime lit room, the standard picture settings of most TVs are too bright for best picture quality.

That super-bright mode will also shorten your HDTV's lifespan (or at least the lifespan of your light bulb, if you have a microprojector system) — which really means something if you have a $10,000 plasma that you were planning on using for 10 or 15 years! Even if you only have to change a bulb (and not replace a tube or buy an entirely new TV), you'll still be out a couple of hundred dollars.

Getting Your Settings Right

In order to *calibrate* (properly configure) your HDTV's picture, you'll need to delve into the TV's menu system, using your remote control and the *on-screen display* (or OSD) that your HDTV provides.

Every HDTV differs, but you'll most likely end up in a menu called "Video Settings" to adjust the picture elements we're about to discuss.

We probably don't own the same HDTV as you do. We've seen/used/ played with many different sets — enough to recognize that there's no way we're going 100 percent match up to the terms that *your HDTV* uses in its menu system for picture quality settings. So keep that in mind — we're using the generic terms here.

Here are the most common picture settings:

- **Contrast (white level):** The control adjusts the *white level,* or the amount of whiteness your screen displays:

  - If your white level is too high, white areas of your picture tend to bleed over into the darker areas that surround them.

  - If your white level is too low, whites don't appear true white.

  Whites and blacks are measured on a scale called IRE (Institute of Radio Engineers) units. These are percentages between 0 and 100 — 0 percent is black; 100 percent is white.

- **Brightness (black level):** How's this for confusing? Your HDTV's *brightness* control adjusts the *black level* that you see on the screen. Seems a bit counterintuitive, doesn't it? If the black level is set incorrectly, dark scenes on your HDTV will be indistinguishable — you won't see the bad guy in the black suit hiding in the shadows.

✔ **Sharpness:** Ever see a fuzzy picture on your screen? The sharpness control is the likely cause — it adjusts the *fine detail* of the picture.

- If the sharpness is set too high, your picture appears *edgy,* often with "blobs" around the edges of objects, instead of clearly defined lines.

- If the sharpness is set too low, you have a fuzzy picture. (Aha, so *that's* why it's fuzzy!)

✔ Two adjustments set the balance of colors:

- **Color:** The color setting is used to adjust the *intensity* of the TV's display of colors — if this is set too low, you'll see only black and white (or grayscale); set it too high, and your colors will bleed together.

  If your color setting is too low, images begin to appear as black and white. If the color level is too high, images take on a reddish tinge — everyone resembles Bozo the Clown!

- **Tint (hue):** On most TVs, this control is labeled *tint,* but a few are technically correct and call it *hue.* This control adjusts your display's color only within the range between red and green.

  This is pretty hard to adjust with your eyes alone — we recommend the calibration systems in the next section to get it done right.

Many HDTVs come with some picture pre-settings — mix-and-match combinations of the settings listed here — designed for specific purposes. For example, Sony HDTVs have settings such as "Vivid" (which is that showroom "torch" setting), and "Pro" (which often is pretty close to being well calibrated for viewing movies in a dark room).

Getting Calibrated

All those video settings discussed in the previous section can be a bit difficult (we think impossible) to set up properly by just "eye-balling it." Unless you have the visual equivalent of perfect pitch, how can you know when the whites are white enough and the blacks are black enough (not to mention whether the reds are red enough)? You can either do it yourself or bring in the professionals.

Doing it yourself

Calibration DVDs and tapes contain a specially designed series of tests and test patterns that you can follow (along with on-screen directions) to get the settings right on your HDTV. Calibration systems provide all the information you need to tune up your HDTV.

If you really have to adjust your video without the aid of a calibration system, here's what we recommend (at minimum):

- ✔ Turn down the brightness and contrast until the levels are about ⅓ to ½ all across your screen.
- ✔ Substantially turn down the sharpness control.

Commercial DVDs

Most calibration systems come in the form of DVDs — which means you won't be using a high-definition picture to tune up your HDTV. Yes, this is a less-than-great situation, but the settings are fairly universal, so an HDTV that's been well calibrated for a DVD input will also look good for displaying HDTV.

High-definition DVDs will be here soon (see Chapter 11). Expect to see high-def versions of the DVDs mentioned here before long.

There are popular calibration DVDs on the market from a couple of sources. Any of these discs will do a great job helping you get your HDTV tuned up and calibrated. Each of these discs includes

- ✔ A funky blue filter — they usually look like those crazy paper 3D glasses you needed to wear for *Jaws 3!* The blue filter lets you properly set up your color and hue settings.
- ✔ Audio-calibration tools that will help you get your surround-sound audio system properly configured (if you have one — see Chapter 19 to learn about surround sound).

You really can't go wrong with any of these discs:

- ✔ Ovation Software (www.ovationsw.com) distributes a couple of calibration DVDs with a series of easy-to-follow on-screen test patterns and signals that let you correctly adjust your TV's picture settings:
  - • *Sound and Vision* **Home Theater Tune-up:** Produced in conjunction with *Sound and Vision* magazine (one of our favorites). It's about $25.

- **AVIA: Guide to Home Theater:** This disc contains a ton of great background material about TVs and home theater. It's about $50.

AVIA Professional is a *seven*-DVD set that (as the name implies) is designed for professional calibrators (see the section titled "Bringing in the pros" for more on this).

✓ **Digital Video Essentials (DVE):** Found online at www.videoessentials.com, this is the definitive calibration disc. It costs about $50. Of all the discs available, DVE has the most tests and the most calibration settings.

One really cool thing about Video Essentials is the inclusion of video footage that you can watch to see the results of your adjustments with actual video, instead of just a test pattern.

THX DVD

If you don't want to spend the money on a calibration disc, you can get much of the functionality without spending an extra penny. Just pick up a movie DVD with the *THX Optimizer* on board. Like the commercial calibration discs, the Optimizer walks you through a series of steps to adjust your display (and your audio system).

Any THX-certified DVD (there were 361 such titles as we wrote this in the summer of 2004) includes the Optimizer.

To get all you can from the Optimizer, you need a blue filter. You can get one from THX by filling out the order form online (there's a link on the front page of www.thx.com. All you need to pay is 2 bucks for shipping/handling.

High-definition D-VHS

The Video Essentials folks (www.videoessentials.com) have responded to the HDTV market by coming up with D-VHS videocassettes that let you calibrate your HDTV with a true HDTV picture, instead of an NTSC-quality DVD picture. These tapes are called *Digital Video Essentials — High Definition*.

You need a D-VHS VCR to use one of these tapes!

Digital Video Essentials — High Definition is available in two resolutions. Either costs about $90. Choose the version that's closest to your TV's native resolution:

- ✓ If you have a direct view CRT or a CRT projection HDTV system, you probably need the *1080i* version.

- ✓ If your HDTV doesn't have a CRT-based display, you probably need the *720p* version.

Bringing in the pros

The absolute best way to get your HDTV properly tuned up is to hire a pro to come to your home and do a professional calibration. Two things separate a professional calibration from the one you perform for yourself at home:

- ✔ **Training:** Your calibration professional should have extensive training and certification, along with a chunk of real-world experience with finicky TVs (things you probably don't have!).

- ✔ **Equipment:** Your calibrator will also have some expensive equipment that can measure the color and brightness of images on your TV screen — this gives a much more precise calibration than using your eyeballs.

A professional calibration usually costs between $200 to $500, depending on what type of display you have.

If you have a professional calibration done, make sure you choose someone who has been certified and trained by the *Imaging Science Foundation* or ISF. The ISF has trained (and continues to train) thousands of home-theater dealers in the art of system calibration. You can find a trained calibration professional near you by searching on ISF's Web site, at www.imagingscience.com.

Don't call ISF directly to ask them to calibrate your system. They'd know how, but that's not what they do for a living. Their Web site includes a searchable (by state) listing of the dealers they've trained to perform this service.

Chapter 6

Magic Black Boxes

· ·

In This Chapter

▶ Playing the resolution game

▶ Understanding scalers and video processors

▶ Doubling or quadrupling your fun

▶ Deciding to go external

· ·

Throughout *HDTV For Dummies,* we've talked about resolution (the number of picture elements that make up a video signal). Analog TV (NTSC) signals have a particular resolution, as do DVDs, and the various types of VHS videocassettes (VHS, S-VHS and D-VHS). And that's to say nothing about the ATSC digital-TV standard, which has 18 (count 'em, 18) different resolutions.

HDTV hardware also has a resolution angle. All HDTVs have one (or a couple) of preferred resolutions at which they will display TV signals. Many HDTVs are what's known as *fixed-pixel* displays — they can show TV signals (no matter what resolution they are recorded in) at a single resolution on-screen.

In this chapter, we explain how all these different resolutions of software (TV programming) and hardware (HDTVs) work together. We also explain the sophisticated devices that make this cooperation happen. Finally, we talk about how you might choose a scaler (you don't *have* to — your HDTV will have one inside it already — but some folks like to go above and beyond).

What the Heck's a Video Processor?

Sometimes the stars align and everything just works. For example, you may tune in to an HDTV broadcast of *Law and Order: Special Victims Unit* (which is broadcast as a 1080i signal) with your Samsung

CRT HDTV (which displays HDTV signals in a 1080i mode). Well, lucky you — the input resolution (the 1080i TV program) and the display resolution (your 1080i CRT TV) match. All is well in the world, and good TV is watched.

But what if you had a Sony Grand Wega LCD rear-projection TV? Those big-screen beauties display HDTV at 720p, not at 1080i!

This TV displays HDTV at 720p because the LCD (like the plasma, the DLP, and the LCoS) is a fixed-pixel display. In the case of the Sony Grand Wega, the LCDs that make the display work (see Chapter 22 for more info) have a resolution of 1386 x 788 pixels — roughly equivalent to the 1080 x 720 of 720p HDTV.

Well what happens in this case (1080i signal into a 720p display) is that a brainy bundle of chips and software called a *scaler* or *video processor* gets involved. A scaler converts one resolution to another, using various mathematical techniques to interpolate what the video signal would look like at a different resolution.

Going up or going down?

If a scaler is converting a lower resolution to a higher one, the process is called *upconversion*. The opposite process is called, unsurprisingly, *downconversion*.

Sometimes downconversion is called *down-resing* because the resolution is moved downward. This term is usually applied in a negative way — for example, when an onerous copy-protection system downconverts an HDTV signal to standard-definition simply because you don't have the right copy-protection software on *all* your components.

The main job of a video-processing system is to up- or downconvert an incoming signal to a different resolution. There are two main benefits of this process for a scaler working with your HDTV:

- ✔ Video signals are matched to the best display resolution for a particular model of HDTV, regardless of the signal's original resolution.

- ✔ Standard-definition video signals are upconverted to a higher resolution to look *closer* to HDTV than standard-definition.

Don't buy anyone's marketing spiel or sales pitch telling you that his or her scaler will make any TV source into HDTV. Good scalers can produce something very pleasing to the eye, which is close to HDTV. But you can't create something from nothing — real HDTV signals can have at least six times as many pixels as standard-definition. In other words, in an upconverted video stream, five out of every six pixels could be "made up" by the scaler. Even the best scaler can't create something as good as the original recording!

Video processors may also perform other tasks — such as 3:2 pull-down processing (discussed in Chapter 21), which lets you watch material based on 24-frame-per-second film properly on a 30- or 60-frame-per-second HDTV display.

De-interlacing your video

The simplest video processors — which actually pre-date HDTV — are those devices known as *line doublers*. Line doublers have been used in high-end home theaters for years, in conjunction with fancy (and expensive) front-projection systems.

Line doublers are also sometimes called *deinterlacers*.

The job of a line doubler is pretty simple — it scales the video by converting an analog (480i) video signal into a progressive scan (480p). By doing this simple trick (effectively a doubling of the scan lines in a CRT TV — hence the name), a line doubler greatly smooths out the picture, and reduces the subtle flicker you normally get from an interlaced picture.

A line doubler works by saving up both of the *fields* in an analog TV signal *frame,* and displaying them twice in a ⅟₃₀-second time slice. (In analog TV, half of the picture is transmitted every ⅟₆₀ of a second, which is a *field;* the whole image is called a *frame.* Chapter 21 covers fields and frames.)

You also hear about devices call *line quadruplers,* which not only deinterlace video, but also interpolate *in between* the lines, to create a picture that's the equivalent of 960p (twice the resolution of 480p). Line quadruplers are really high-end devices for use with the most expensive ($40,000 and up) CRT front-projection HDTVs.

Getting fancy with scaling

The advent of fixed-pixel HDTV displays — plasmas, LCDs, DLPs and LCoS — created a need for something beyond just a simple line doubler (not that a line doubler is all that simple!). Many fixed-pixel displays actually have rather funky native resolutions that require *all* incoming video signals to be scaled.

For example, that Sony Grand Wega TV we used as an example earlier doesn't have a native resolution of *exactly* 720p (1280 x 720 pixels); it actually has a resolution of 1386 x 788: Every signal, even 720p signals, must be converted to the higher resolution!

This demand has led to the development of scalers that convert *any* incoming video signal (well, any *standard* incoming video signal) into a specified output resolution.

Choosing Scalers

We don't want to make you think about scalers too much. If your HDTV needs a scaler to operate (and most need at least a deinterlacer/line doubler), then it already has one built in.

Let's repeat that: If your HDTV needs a scaler, that scaler should be in there already. You don't *need* to spend even *more* money!

There are, however, some limitations to internal scalers. Sometimes an *external* scaler offers advantages. Here are three scenarios that may call for an external scaler:

✔ Some HDTVs have internal scalers that accept signals of only certain resolutions. For example, some HDTVs may accept only 480i, 480p and 1080i inputs. If your external HDTV tuner puts out only a 720p signal (admittedly rare), you're out of luck.

✔ Some HDTVs (CRT tube-based TVs, either direct-view or projection systems, usually) contain an internal scanner that deinterlaces analog video sources (that is, converts 480i to 480p) but doesn't upconvert analog signals to a higher resolution. There's nothing wrong with this — 480p will probably look better than analog's ever looked! — but the picture won't be using the full capacity of your HDTV.

✔ You may just be a person who wants the best! The scalers built into most HDTVs are good enough for most owners. But, if you demand those last few percentage points of picture perfection from your system, you may want a fancier external scaler.

Just to give you an idea of what an external scaler is all about, we mention a popular (and well-reviewed) model: the DVDO iScan HD. This scaler (which retails for $1,499) can convert any incoming video signal to any resolution between 480p and 1080p (yes, we said, 1080*p* — that is one of the 18 ATSC standards, but it isn't commonly used because almost no HDTVs can display that high a resolution). And, if your display has a funky native resolution, that's no problem — the iScan HD can be programmed for custom resolutions, too!

The iScan HD also does a bunch of other cool stuff, too:

✔ It acts as a video-switching hub, so you can run all of your source devices through it. (See Chapter 19 for more on video switching.)

✔ It switches and routes analog *and* digital audio signals as well.

✔ It has a special circuit "Precision AV LipSync" that makes sure your video and audio are perfectly matched up — even if the original broadcast isn't! No more movies that look like poorly dubbed Kung Fu flicks.

✔ It includes a computer interface, so you can download software upgrades over the Internet to keep your iScan HD up to date.

You can find out more about the iScan HD at the following URL:

```
www.dvdo.com/pro/pro_ishd.html
```

Part III
HDTV Channels

The 5th Wave   By Rich Tennant

"Darn it, I wish they'd sent one that already has the HDTV tuner built in."

In this part . . .

*I*f you haven't figured it out yet, we're kind of obsessed. See, we like to just sit and look at A/V equipment even when it's turned off. We just love looking at turned-off HDTVs, especially the backs where all the many ports and connectors are located.

Normal people, like you, probably want to actually look at an HDTV program, not an inert HDTV. OK, so we can't fault you there. A couple of years ago, however, we would have shaken our heads sadly and told you there were none to be found. Today, however, there's an ever-increasing amount of HDTV programming available — programming that will elicit the appropriate *oohs* and *ahhs* from people visiting your house.

In this part, we expose you to various ways you can get high-definition content onto your HDTV. We start with a general look at the signals available from the over-the-air broadcasters, cable companies, and satellite companies. We reveal why not all signals look the same from these sources — even if they're broadcasting the same movie. Different sources have different compression and transmission capabilities. They all look great. But some will look greater than others.

Then we drill down on each of the three major originators of HDTV signals, starting with the over-the-air broadcasters. We can thank the FCC for prodding the TV stations in the direction of HDTV, but really thank the HDTV manufacturers for making displays inexpensive enough to drive HDTVs into lots of homes — giving the TV stations rationale to move to more digital programming. We tell you probably more than you want to know about HDTV antennae, and how to optimize your connections to the OTA (over-the-air) folks.

Then we do the same with the cable industry, looking at how the cable firms are positioning to deliver high-definition cable programming. We tell you about the way HDTV signals are encoded and decoded, and the role of your analog and digital set top boxes. We also tell you about the new CableCARD capability that you are going to start seeing in HDTV devices.

We wrap up the HDTV programming discussion with a look at satellite programming firms and what they can offer. In addition to talking about DirecTV and DISH (the two big guys), we also explore the latest player in satellite-based programming, VOOM. We walk you through what's available, how to upgrade to HD satellite service if you're already a customer, and even how to combine satellite service with Internet access. Whew, that's a lot! Aren't you glad you came here first!

Chapter 7

Who's Showing HDTV?

*I*n the vein of "all dressed up and nowhere to go," HDTVs would be no good without content to show on them — native, high-definition content, that is. Hollywood has done its usual: "If there's a place to put the content, we'll come out with it." And the manufacturers have said, "If you make the content, we'll build the systems." Lucky for us, DVDs and higher-definition camcorders have created a lot of reasons to want higher-resolution TVs, and that has prompted the broadcasters to follow suit with their own HDTV content.

Broadcast TV stations, for example, are heeding the FCC (Federal Communications Commission) and beginning HDTV broadcasts. Cable companies are offering at least some HDTV programming (and usually not just "some" programming but "a lot") in most of their major markets. And satellite providers are offering HDTV to pretty much the entire United States — so almost no one is outside the "footprint" of HDTV these days.

In this chapter, we take a high-level look at who's offering HDTV, what it takes to get the service, and what *kind* of HDTV content you can get from the different providers. That last item can be important to know if, for example, you bought your HDTV specifically to watch *The Final Four* (we predict a big Duke run next year!) or for seeing Tony and Carmela in the utmost detail on *The Sopranos*.

In the three chapters that follow in this part of the book, we get into a lot more detail about broadcast, cable, and satellite HDTV — the programming, its availability, and the equipment requirements. Consider those discussions in this chapter a quick, up-front primer — move forward in the book when you're looking for the details.

Looking at Who Has HDTV

We're in the midst of HDTV-mania. If you happened to have a dead cat handy (and *no*, we're not cat haters! We love cats, and don't want them dead, unless they've been very, very bad kitties), you wouldn't be able to swing it very far before you hit an ad for HDTV sets, HDTV programming, HDTV services from a cable company, or *something* HDTV.

This situation makes us happy. We've been going to the CES (the Consumer Electronics Show — the biggest HDTV-related trade show there is) every year for a long time. And every year we hear, "This year, *this year,* is the Year of HDTV!" Well, we've finally reached that point — in what must be the fifth or sixth official "Year of HDTV" (YOH for short). Thank goodness.

The reason we know that it's now YOH is the sheer volume of HDTV programming that's become available from different sources, and the profusion of ways the average person can now access that programming.

Broadcasters

One group of TV providers who have begun to really see the light is the broadcast networks (such as ABC and CBS) and local affiliate stations that broadcast the network content over the airwaves. This hasn't happened purely because the broadcast folks are being good TV citizens. It's because the FCC has mandated a transition from analog to digital TV. Eventually, all of these broadcasters will need to turn off their analog signals and send out DTV broadcasts (note that we said DTV, not HDTV — lower-resolution 480p or even 480i signals can be broadcast).

It's not all pressure from the FCC, however, that is driving this. Broadcast networks are increasingly competing with what we'll call *cable networks* — though it's a bit of a misnomer, we use this term to describe networks you can only get via cable or satellite (such as HBO or TNT). Broadcasting in HDTV — especially for big-ticket

items like prime-time shows and major-league sporting events —
gives the broadcast networks a leg up on cable networks (many of
which are still stuck in Analog TV Land).

Four of the five major broadcast networks (that would be ABC,
CBS, NBC, Fox and *the* WB) are broadcasting at least some HDTV
content. As we write, Fox is the lone holdout, but it promises HDTV
by the time this book hits the shelves.

The best thing about broadcast network HDTV (often called *OTA*,
or over-the-air) is that it's free. Free as in *free beer*. No cost to you
(except maybe having to watch some bad ads). All you need is an
HDTV with a built-in or external tuner, and an antenna. In Chapter 8
we give you some more detail on this.

Cable and satellite networks

The cable networks (the networks you can't get with a rabbit-ear
antenna) have not been sitting around idly while broadcast networks began to send out this free HDTV to their viewers. In fact,
many cable networks have developed and launched their own
high-definition channels — ranging from movie channels (such as
HBO or Showtime High Def) to sports channels like ESPN-HD.

The biggest problem that cable networks have had is finding cable
or satellite networks to carry these channels to customers. To
understand why, skip ahead to the section titled "All HDTV Signals
Are Not Equal." Go ahead — we'll wait right here for a moment.

Basically (in case you *didn't* skip ahead), cable and satellite systems have a limited amount of *bandwidth* (or slots for TV channels)
within their broadcast systems. HDTV uses five to eight times as
much bandwidth per channel as does analog TV — or to reverse
that, you can fit five to eight analog TV channels in the slot occu-
pied by one HDTV channel. So it's taken a bit of time for cable and
satellite companies to begin to show interest in carrying HDTV
channels — they thought they could make more money off eight
analog channels of fly-fishing and underwater basket-weaving than
they could from one HDTV broadcast of *The Final Four*.

The good news is that both groups of TV providers *have* begun to
carry at least a handful of HDTV stations — often more. To get into
these HDTV cable signals, you need the following:

  ✔ An HDTV service contract with your cable or satellite com-
    pany. Unlike broadcast, this isn't free — you've got to pay
    the piper.

✔ An HDTV satellite receiver or set-top box (for cable).

In Chapter 9 we discuss digital-cable-ready TVs and CableCARDs, just now hitting the market, which will let you skip the set-top box.

Specialized HDTV stations

Existing broadcast and cable networks aren't the only ones to realize the potential of HDTV. A small flurry of new networks specifically delivers HDTV channels to cable and satellite providers.

A good example of this is HDNet (www.hd.net), the brainchild of Dallas Mavericks owner Mark Cuban. HDNet was launched with the sole purpose of providing a range of original programming (news, sports, and series) — along with licensed programming from other studios (like Andy Richter's show!) — all in full 1080i HDTV. Danny is particularly interested in HDNet's show "Bikini Destinations" (Pat made that up!). You can get HDNet's two channels on both of the major satellite networks, as well as a growing number of cable provider's networks.

HDNet was the first, but not the last high-def-only network. For example, a cable-only network (owned by a consortium of cable companies) called INHD (www.inhd.com) provides two channels with a variety of original and licensed programming of sports, movies and other content (hey, that sounds familiar, huh?).

Making Your Choice

We have some good news and some bad news. The good news is, you probably have a choice of where you get your HDTV content. The bad news is — well, you have to choose. This means, if you're like us, hours of research and poring over Web sites, trying to figure out what works best to fit your HDTV needs.

Well, there isn't a magic bullet for you here in *HDTV For Dummies*. Because HDTV availability is highly dependent upon *exactly* where you live (we mean right down to the street address — it can differ even within neighborhoods), we can't give anything but the most general advice.

There's nothing wrong with mixing and matching amongst these different sources. For example, if you want local HDTV content along with your satellite-TV source, you need to hook up an antenna to your dish and pick up the OTA broadcasts — luckily, most HDTV satellite receivers have a built-in OTA HDTV tuner, so you don't need extra equipment (beyond the antenna).

So, given that wishy-washy disclaimer (sorry, but it's true!), here's Pat and Danny's official advice:

1. **Figure out what's even available.** In each of the following three chapters, we give you some pointers to Web sites and other resources that will help you find out what you can get in your house.

2. **Look at your budget.** Keep in mind the fact that "free" OTA HDTV may not be free if you have an "HDTV-ready" system, and need to spend hundreds of dollars on an external HDTV tuner. Cable, on the other hand, might include a monthly fee, but doesn't require any "up-front" expenses for tuners or set-top boxes. Many cable companies give you local HDTV channels free for the price of the set-top-box rental ($10/month or less, typically). Satellite may have lower monthly fees than cable, but also requires an up-front purchase of the receiver.

3. **Examine closely the channel lineups available to you.** Remember that quantity and quality are two different things. For example, the new HDTV satellite company Voom (www.voom.com, discussed in Chapter 10) offers 35 HDTV channels — more than anyone else — but you may not be interested in watching all of them. For example (and this is a totally made-up example, so please don't sue us, Voom, because we love you!), do you really want to spend your HDTV viewing time watching some nature special on the mating habits of frogs? Look for the channels you love.

4. **Consider the performance.** In the following section "All HDTV Signals Are Not Equal," we discuss some of the ways that various TV providers throttle back their HDTV signals to save bandwidth on their networks.

In the end, you may find something that we haven't mentioned, or something less than coldly rational and logical (Mr. Spock to the bridge!) that makes you decide on an HDTV provider. For example, for Pat (actually, for his wife, who is a *huge* baseball fan) it was a no-brainer — he wanted satellite, but then his local cable company started broadcasting 100+ San Diego Padres games in HDTV each year. Case closed — cable it is for Pat.

All HDTV Signals Are Not Equal

Not all HDTV signals are equal across all providers. When the signal is sent from its source to you, it is usually compressed to cut costs and bandwidth requirements. How much the signal is compressed — and what digital encoding scheme is used — determines a lot about what you see on your HDTV.

MPEG-2 is the standard used by digital-TV broadcasters today to compress, encode, and then ultimately decode the TV programs. This is necessary because there is not an unlimited amount of space available on the cable, satellite, and broadcast-TV networks for transmitting these signals. (See Table 7-1, "Who Has the Most Room for HDTV?")

Table 7-1: Who Has the Most Room for HDTV?

	Terrestrial	*Satellite*	*Satellite*	*Cable*	*Cable*
Bandwidth	6 MHz	24 MHz	36 MHz	6 MHz	6 MHz
Modulation	8 VSB	QPSK	QPSK	64 QAM	256 QAM
Bit Rate	19.39 Mbps	27 Mbps	40.44 Mbps	27.7 Mbps	38.8 Mbps

Source: Ultimate AV Magazine (June 2004)

So suppose an over-the-air broadcast TV signal starts out its life at the central network hub as a 1920-by-1080i signal. This signal can first be encoded for broadcast at a rate of around 995 Mbps. By the time it is sent to you over the air, it's compressed to a mere 18Mbps signal. That's a lot of compression, but the picture still looks great.

If the station chooses to compress that signal a little more, say down to 13 or 14 Mbps to make room for other channels in the same signal, then the compression is even greater. We can tell you that the difference between an 18Mbps and a 14Mbps signal is indeed noticeable, especially when the programming contains a lot of motion. And don't forget, once the signal enters your HDTV set, it will probably encounter *another* round of downward resolution as it tries to put the original 1080i image on, say, a 720p TV set.

Satellite providers also have similar compression challenges. Transponder space (the satellite-located transmission systems) is so expensive that compression is required for everything. Popular channels are typically encoded at 15 Mbps or more before they go to the satellite operators. Depending on the available satellite bandwidth (could be 24 MHz or 36 MHz, which yields 27 Mbps and 40.44 Mbps of bandwidth respectively), satellite operators can compress these 15Mbps signals down to 13.5 Mbps, so they can cram two or three HDTV streams on these satellite signals, respectively.

Cable operators have about the same options with the two major digital-cable modulation schemes in use today — 64QAM, which offers a max bandwidth of 27.7 Mbps, and 256QAM, which has 38.8 Mbps available. They can compress the signals further, and cram more channels into each signal, or they can offer higher-quality signals as a way of competing better in the market. After all, the 38.8Mbps rate of the cable companies is twice the bandwidth per signal compared to the 19.4Mbps data rate of the broadcast TV folks — so they can send two HDTV streams for each single HDTV stream that the over-the-air stations can send.

Cable companies have to deal with the sheer number of places where the signals are converted from digital to analog and back. Signals start out as digital, but may be converted to analog composite by the distributor. The signal is converted to a *digital composite* (that is, component) signal by the recipient cable operator. Then, as it's fed into the TV display unit, the signal is converted to one of several types of video signal: analog component, S-video, or composite. TVs that take in digital signals via digital interfaces can reduce some of this conversion — as can improvements in signal distribution between the broadcasters and the operators.

So, as complicated as this is, what's it all mean for the average buyer? Here's how we can net it out for you:

- ✔ Not all HDTV signals are going to be the same. It depends on how the broadcasters compress the signals for the stations that you want to watch. Channels with a lot of action (like sports) tend to get more bandwidth; channels with a lot of static pixels — like the Home Shopping Network or any cooking channel — can survive more compression.

- ✔ If you plan on watching a few channels a lot, do some research. Your ultimate goal would be to find out how much bandwidth these channels are using. You can't (easily) measure this yourself (and it's about impossible without installing the service, if you're using cable or satellite), but try to find a local bulletin board or newsgroup online to see what the local HDTV nerds have to say. For example, Pat spends a lot of time on http://hdtv.forsandiego.com/, reading about experiences with local HDTV options. You'll find enthusiastic, smart folks who have the tools and the know-how to examine HDTV bit streams, measure them, and come out with concrete evaluations. Also, watch your neighbors' TVs now and then, and compare what you can — if it *looks* different, it probably is.

Who's required to do what?

Your over-the-air digital TV broadcaster only has 6 MHz (19.39 Mbps) of bandwidth over which it can send its TV signals. It can fill this in anyway it chooses, combining such broadcasts as HDTV, SDTV, weather images, and even FM broadcasts onto one bandwidth stream. The FCC, through its DTV rules, has said broadcasters must transmit one standard-definition digital-TV signal — 480i — and has not said anything about requiring HDTV.

So TV stations have to make some economic decisions about how best to use that signal. In one chunk of 6MHz bandwidth, a broadcaster could send one full-quality, 720p 60-frame-per-second (fps) HDTV program; or, two shows, one (say) for a 720p or 1080i HDTV show and one 480i SDTV show, both at 30 fps; or, four 480i SDTV 30 fps shows. How a station mixes and matches its signals depends as much on its technology as its positioning in the market.

Chapter 8

Something's in the Air

*1*n most ways, HDTV is an amazing leap into the future. It's digital! It's high-definition! It's shiny and new! It's expensive!

But HDTV does have one aspect that brings us all back to the olden days of black-and-white TVs — the antenna has reappeared! And with the antenna comes all the fun of Dad standing on one foot and twisting the rabbit ears *just so* to bring in that hard-to-tune channel.

Okay, so it's not that bad (most of the time), but it can be challenging to tune into over-the-air (OTA) broadcast TV. We're not going to kid you here — cable-TV or satellite setups are much easier to use than an antenna.

But OTA HDTV is free. FREE! And nearly every household in America is in range of at least one of these free HDTV stations (99 percent of homes, according to some sources).

In this chapter, we discuss how you can figure out what OTA HDTV signals are available in your location (there are some great online tools!). Then we help you pick out the equipment you need (an antenna and a tuner), and finally we tell you how to get it all set up.

Finding Local HD Broadcasts

Before you do *anything* else, do your homework. In the case of OTA HDTV, this involves simply spending a little bit of time online looking for broadcast HDTV channels. Now you don't *have* to do this research online, but if you have Internet access, that's the best way to do so.

If you don't have Internet access, talk to the retailers from whom you're buying your HDTV — they probably have firsthand knowledge (or at least anecdotal information) about local HDTV signal availability. They may also have a kiosk that provides access to the same online information we're going to talk about shortly.

The Consumer Electronics Association (CEA) is the huge (and hugely influential) trade organization that includes just about every HDTV manufacturer in the world. So it has a vested interest in getting people to buy HDTVs. This interest is sometimes manifested as lobbying efforts with the FCC, or efforts to develop marketing and industry. It also pops its head up in the form direct-to-the-consumer education efforts.

In the case of HDTV, the CEA had the brilliant idea (we're not being facetious, it really *was* brilliant!) of providing an extremely easy-to-use HDTV "signal finder." In conjunction, with a company called Decisionmark, the CEA has created this online system (called TitanTV) that lets you enter your address information, press a button, and come out the other side with a nice listing of all your HDTV-channel choices.

Just go to `www.titantv.com` and follow the on-screen instructions. You can also access TitanTV at many HDTV retailers, and come home with a printout of your available stations.

Not only does TitanTV find OTA stations for you, it also comes up with the cable and satellite stations available to you. You can do a quick comparison and see what best meets your needs.

The CEA and Decisionmark have also teamed up to provide you with another way to view a listing of local HDTV broadcasters with their AntennaWeb.org system, which we discuss later in this chapter, in the section titled "Choosing the right antenna." AntennaWeb.org is basically the same underlying database, optimized to help you choose an antenna — TitanTV is designed to help you see what shows are being broadcast.

Tuning In

HDTV breaks the old TV paradigm in many ways. One big difference is that not all HDTVs have a built-in TV tuner. Back in the olden days of NTSC and analog TV, this was quite rare — a few high-end TVs were "monitors" and required an external tuner, but they were definitely in the minority.

With HDTV and OTA HDTV broadcasts, the situation has been reversed — at least for the time being (more on that soon). Many sets sold as HDTVs today are *HDTV-ready,* but don't have any electronics inside them that can pick up an OTA HDTV broadcast (or *any* ATSC broadcast, whether high-definition or standard-definition — and if you have no idea what we're saying right now, check out Chapter 1 for a definition of ATSC and NTSC).

The government is requiring TV manufacturers to include built-in tuners that can pick up OTA HDTV, starting with bigger (36+-inch) TVs in 2005. Soon this lack of tuners won't be a problem.

Building on a built-in tuner

If you have an HDTV with a built-in ATSC tuner, you're just about all set. All you need to do is find the appropriate antenna (see the following section "Antennas A to Z"), make the connections, and go. It's really that simple — or at least it can be.

Just follow the instructions in your HDTV's manual for tuning in the HDTV stations — we can't help you there, since each HDTV on-screen setup process is different.

Some HDTVs with built-in ATSC tuners also have special tuners that can decode *QAM*-encoded HDTV signals. QAM is the system used by most cable-TV networks. This means you may be able to pick up your local broadcast stations by just plugging in your cable TV connection. Note that this is different from the DCR (digital-cable-ready) systems we discuss in Chapter 9. And no, this has nothing to do with using an OTA antenna, but we thought it was a nice morsel of information to know in case you're ever asked this question at a neighborhood BBQ.

Adding on a tuner

If you own an HDTV, chances are good that you won't have that built-in ATSC tuner we mentioned in the previous section. Your

HDTV probably has an NTSC tuner — which can pick up analog broadcasts — but it probably is only HDTV-ready, so you need to pick up an HDTV tuner box that you can connect between your antenna feed and your HDTV.

The biggest problem with HDTV tuners (whether they're in your TV, or external) has traditionally been an economic one. They are (or at least were) darned expensive. Even a year or two ago, it was common to see HDTV tuners that cost $1,000 or more (this effectively explains why so many HDTVs were sold as HDTV-*ready:* Not everyone with an HDTV uses the OTA tuner, so why drive the cost of an already-expensive HDTV through the roof?).

The good news is that prices have come down. Way down. As we write (summer of 2004), you can buy a name brand (Samsung, in this case) HDTV tuner for about $350. That's less than six months' worth of double espressos at Pat's favorite café! And the prices are going nowhere but down.

When you're choosing an HDTV tuner, here are just a few things to consider:

✔ **Digital outputs:** If it's at all possible, you should use a digital cable connection between your tuner and your HDTV — either DVI-D or HDMI (see Chapter 3) is by far the most common (1394/FireWire used to be common, but it's rare these days). Make sure that the outputs of your tuner match the inputs of your HDTV.

The newer HDMI system is *backward-compatible* with DVI-D with the use of a simple adapter. So you can mix and match DVI-D and HDMI freely.

If your HDTV tuner's DVI-D or HDMI output uses the HDCP copy-protection scheme, make sure the DVI-D or HDMI input on your HDTV does, too. Otherwise the system may down-res the signal, giving you a non-HDTV picture.

✔ **Analog outputs:** While you'll want to use your digital outputs, if possible, it's handy to have a full set of analog outputs on the HDTV tuner, for making connections to other devices (such as a DVR).

✔ **Output resolution:** Most HDTV tuners can be adjusted to match the best resolution for your HDTV. Some HDTVs require a specific signal resolution (such as 1080i); if yours does, make sure your tuner can give you output at that resolution.

✔ **Satellite capability:** Some OTA HDTV tuners also include satellite-TV receivers (for services like DirecTV or DISH Network, see Chapter 10). Well, the satellite companies would flip it around and say their receivers include OTA tuners. Either way, this can be handy if you're using a satellite service for premium HDTV channels (like ESPN-HD and HBO) and using an antenna to pick up local HDTV channels.

The FCC is requiring TV manufacturers to begin including built-in ATSC tuners in their TVs, as part of the overall industry transition to digital TV. By the middle of 2005, all TVs larger than 36 inches will require an on-board ATSC tuner — over time, this requirement will filter down to smaller TVs, and even to devices like VCRs (which have their own NTSC tuners these days).

Antennas A to Z

The other half of the OTA HDTV equation is the antenna. Yep (just as we said at the beginning of the chapter), the rabbit ears are back. Well, you probably won't want to bring back some of those unwieldy old antennas of years past, because the TV antenna has gotten in shape for the new millennium.

Choosing the right antenna

You probably *could* use those old rabbit ears — there's nothing that incredibly unique about picking up HDTV broadcasts — but tuning them in (which is what the HDTV tuner does) is a complex process. Luckily, there are a lot of great new antennas on the market that work particularly well with HDTV broadcasts.

Understanding antenna features

There are a few features of an antenna to consider:

✔ **Location:** Inside the house or outside? Well . . .

- *Outdoor antennas* work better, but they also require work, space, and access outside your home.

- *Indoor antennas* may work well enough for you if you're close enough to the transmission towers.

✔ **Direction:** Where to point the thing? Look here . . .

- *Directional antennas* can be aimed toward a specific point (or within in a certain angular vector).

Directional antennas usually are better at picking up distant signals, and can reduce *multipath* distortion (what happens when the signal bounces off buildings and other structures — you receive the signal multiple times).

- *Multidirectional antennas* pick up signals coming from — you guessed it — multiple directions.

Multidirectional antennas can be mounted just about anywhere, and pick up signals from widely dispersed transmission towers.

✔ **Amplification:** Some antennas have a built-in *amplifier* that increases the strength of weak broadcast signals before they get to your tuner.

Amplification is generally a good thing, unless you are close to the broadcast antenna — you don't want to *overamplify* a signal.

Making the right choice

It's really difficult for even a technically savvy person to determine which antenna is right for a particular home — there's just a swarm of variables to figure out.

That's why we're so happy that the CEA and Decisionmark (the folks we mentioned back in the first part of this chapter) have come up with the AntennaWeb.org system (that is also the URL: http://antennaweb.org).

AntennaWeb.org is a Web-based database that "knows" your HDTV situation from your address, the location of HDTV broadcast antennas near you, and some sophisticated software that models HDTV signal propagation in your area (by looking at various geographical and demographic data — such as the terrain and the presence of high-rise buildings). Just go to the Web site, enter your address, answer a few questions about your surroundings and click the Submit button.

Up will come up a result like the page shown in Figure 8-1. The AntennaWeb.org system uses a graphical representation (a pie chart/color-code system) called the *antenna mark* to identify seven different antenna types with unique different colors and coverage "zones." The pie chart is sort of like a Trivial Pursuit game piece — the more slices of the pie you have, the better. The coverage zones represent the footprint of the antenna based on its designs and use. All the HDTV broadcasts in your area that you can at least theoretically pick up will be shown (as well as the analog stations), and each will be color-coded.

To find the right antenna for you, simply note the color code for the stations you want to pick up, and go shopping for an antenna that matches that color.

Please note how they fill in the antenna-mark pie chart — clockwise, starting with the yellow around to the pink. Most antennas work for all the "lesser" color codes. It's common to find high-power directional antennas that also work fine as short-to-medium-range multidirectional antennas. So if your listed signal sources require antennae coded yellow and green (the second color after yellow in the scheme), you can buy a green-code antenna — it will also pick up the yellow-coded broadcast signals.

Installing your HDTV antenna

When you've picked out your HDTV antenna, there's the simple matter of installing it. If you've selected an outdoor antenna (our recommendation for most users), you need to mount it.

First of all, follow the directions included with the antenna — particularly any safety directions regarding such installation procedures as grounding and wiring. We want you to be safe, and *not* to burn your house to the ground for the sake of HDTV (not even we are that fanatical about HDTV).

AntennaWeb - Microsoft Internet Explorer

File Edit View Favorites Tools Help

Back · Search · Favorites · Media

Stations

Address Location Results
The address you entered was located at the **Street** level.

Stations and Antenna Types
TV reception is determined by the size and type of antenna and the direction in which it is pointed. To determine the right antenna, use the color coded Antenna Type to select the channels that you wish to view. Antenna types are color coded according to the size and type of antenna needed for reception. The list is arranged in order of ease of reception, with the stations requiring the smallest multidirectional antenna at the top, to those requiring the largest directional antennas at the bottom.

Use the "Compass Orientation" listed below to point your antenna or rotor system for the channels you wish to receive. Please note that "Compass Orientation" is referenced to magnetic North.

View Street Level Map

⊙ Show All Stations   ○ Show Digital Stations Only   ○ Show Analog Stations Only

DTV	Antenna Type	Call Sign	Channel	Network	City	State	Live Date	Compass Orientation	Miles From	Frequency Assignment
*	yellow - uhf	KSWB-DT	69.1	WB	San Diego	CA		138°	17.3	19
*	yellow - uhf	KGTV-DT	10.1	ABC	San Diego	CA		228°	11.2	25
*	yellow - uhf	KPBS-DT	15.1	PBS	San Diego	CA		137°	17.2	30
*	yellow - uhf	KFMB-DT	8.1	CBS	San Diego	CA		228°	11.3	55
	green - uhf	KNSD	39	NBC	SAN DIEGO	CA		138°	17.3	39
*	green - uhf	KNSD-DT	39.1	NBC	San Diego	CA		138°	17.3	40
	green - uhf	KUSI	51	IND	SAN DIEGO	CA		137°	17.3	51
	green - uhf	KSWB	69	WB	SAN DIEGO	CA		138°	17.3	69
	green - uhf	KPBS	15	PBS	SAN DIEGO	CA		137°	17.2	15
	green - uhf	KGTV	10	ABC	SAN DIEGO	CA		228°	11.2	10
	green - uhf	KFMB	8	CBS	SAN DIEGO	CA		228°	11.3	8
	blue - uhf	KSDX-LP	29	IND	SAN DIEGO	CA		137°	17.2	29
*	blue - uhf	KUSI-DT	18	IND	SAN DIEGO	CA		138°	17.3	18
	blue - vhf	XETV	6	FOX	TIJUANA	MX		162°	28.7	6

Done   Internet

Figure 8-1: Choosing an antenna at AntennaWeb.org.

If you've chosen a directional antenna, part of the mounting process will be aiming the antenna. The AntennaWeb.org results page actually includes (as you may have noticed in Figure 8-1) a compass bearing from your house to the antenna towers. So you can pull out your handy-dandy Boy or Girl Scout compass and do some aiming.

You can also use a *signal-strength meter* — many HDTVs or HDTV tuners have one built in — to determine how small adjustments in the antenna's position or direction affect your channels. Here's where it's handy to have a set of FRS (Family Radio Service) walkie-talkies, so you can communicate with the person in the living room who's viewing the meter while you're on the roof!

Tuning into HDTV stations is a truly digital experience — either it works or it doesn't. For the most part, there's none of the fuzzy, half-visible pictures that you might remember from the old antenna days of analog TV. With the proper antenna, properly aimed, you might pick up all your local channels right away.

Or you might not — HDTV broadcasts are still a work-in-progress in many parts of the country. Sometimes it can be a hit-or-miss proposition — stations that *should* work don't, or only work part of the time.

This situation is getting better every day, as broadcasters fine-tune their systems, and as HDTV tuners become more sophisticated and adept at tuning in stations. As HDTV becomes more popular, many broadcasters are also turning up the power on their transmitters (so far they've been keeping the broadcast power down to save money!).

Chapter 9

The Cable Guy

*M*ost Americans get their TV (NTSC or ATSC) from their local cable company — something on the order of two-thirds of all households subscribe to cable-TV services. Cable companies (generally referred to as *MSOs* or *multiple-system operators,* as most cable companies own and operate cable systems in dozens — or hundreds — of different cities) have been extremely aggressive over the past ten years, rolling out new services like digital cable, cable modems, voice services (cable telephony), and Video on Demand (VoD).

Cable companies have, however, taken their sweet time in the HDTV realm — many cable companies decided to drag their feet (for various technical and business reasons) when it came to offering HDTV to their customers. Luckily, the threat of HDTV from satellite — combined with technical advances and some regulatory agreements — has given cable TV providers a big burst of HDTV energy. Now we're even seeing ads from our local cable companies bragging about how they have more local HDTV than any satellite company.

In this chapter, we give you the basics on how to get HDTV from your cable company, and what to expect when you do. We also tell you about some of the other services you can get from your cable company (things like digital cable, which are actually the majority of channels available on your HDTV, even though they aren't HDTV themselves). Finally, we talk about an exciting new trend — a new system of TVs and "smart cards" that let you forgo the cable box and plug directly into the cable coming out of your wall.

High-Definition Cable

The crème de la crème of cable services is HDTV over cable, and it's nearly everywhere. Cable companies are now offering HDTV services in 83 of the 100 largest markets in the U.S., according to the National Cable & Telecommunications Association, or NCTA (www.ncta.com), the largest industry group of cable operators.

Getting on the QAM bandwagon

The biggest difference between HDTV signals on a cable system and those broadcast over the air (see Chapter 8) revolves around the system used to *modulate* an HDTV signal for transmission as radio-frequency signals across the cable system.

Broadcast HDTV uses a system called 8-VSB (vestigial sideband), while cable usually uses a system called QAM (quadrature amplitude modulation). What's the difference? Well the technical specs are for the engineers — what's important to HDTV viewers is that you need a different kind of HDTV tuner to receive QAM signals than you do to receive 8-VSB (though many newer TVs are now including a tuner that can tune into both types of signals).

For most people, HDTV cable service will come through an HDTV *set-top box* (or cable box) that has a built-in QAM tuner.

Encrypting and decrypting

The other big difference between broadcast and cable HDTV concerns *encryption* and "premium" cable channels. Broadcast HDTV channels — over-the-air broadcasts that you pick up with an antenna — are free for the picking (up). If you receive the channel, you can watch it.

Many cable TV HDTV channels, on the other hand, are considered premium or for-pay content — you have to pay extra every month to receive them. Because of this arrangement, most cable systems use an encryption system that scrambles the signal (using an encryption *algorithm* or formula). This means that even if you have a QAM tuner (whether built into your HDTV or attached to it), you won't be able to watch these programs without some hardware to *decode the encryption* (in English translated from Geekese, that's *unscramble the picture*).

In some cable systems, the local broadcast HDTV channels are transmitted across the cable system unencrypted — if you have the appropriate tuner, you can view these channels without some sort of extra decryption hardware.

While a very big exception to this rule is starting to hit the market, the current situation for most cable HDTV viewers is this: You need a set-top box from your cable company to view HDTV. This set-top box handles the QAM tuner duty, connects to your cable company to authorize you as a paying customer, and does the decryption and unscrambling of any premium channels. Figure 9-1 shows a typical HDTV cable set-top box.

Figure 9-1: Pace's DC-551 HD is the kind of set-top box your cable company may provide.

Pricing for HDTV set-top boxes varies widely by cable company. Some may charge you a monthly rental for the set-top box, but nothing for your HDTV channels. Others may not charge for the box rental, but then charge you on a per-channel or package basis for HDTV channels — others may have a combination of these two approaches. In many cases, you're required to get the digital-cable packages in order to get HDTV channels beyond ABC, CBS and NBC. (Frankly, it's a mystery to *us* what *your* cable company offers, pricewise, so check out its Web site.)

Your HDTV cable box will typically also control and enable all the cool (but non-HDTV) digital cable functions that you'll want to use on your HDTV — things like VoD and on-screen program guides.

We are writing this just as a midsummer 2004 deadline is looming that requires cable companies to provide services to customers *without* a set-top box. See the section at the end of this chapter titled "Going Boxless" for more details on this new development.

Digital Cable

The next notch down (picturewise) from cable HDTV is *digital cable*. This name has been the source of almost boundless confusion among many smart people we know — simply because the whole "digital" thing gets people to thinking of DTV — and once you have *D*TV in your head, well then, *H*DTV follows.

So let's get this out of the way right away. Repeat after us: "Digital cable is NOT HDTV, digital cable is NOT HDTV . . ."

Defining digital

There, we got it out of our systems (and hopefully yours). Now, we should hasten to add that in some places, HDTV channels may be part of (or an optional component of) a digital-cable "package." But that's just marketing.

Digital cable itself means two things:

- **Digital transmission and compression:** Standard-definition NTSC analog-TV channels are put through a digital wringer and come out the far end as digital channels (MPEG-2, typically, if you care to know the tech stuff) that take up less space on the cable system (in other words, you can cram more of them onto the cable system, so you can get the mythical 500 channels). These digital channels are then digitally transmitted using a QAM system.

- **Two-way communications:** Digital cable utilizes two-way communications over the cable system (the set-top box can send data "upstream" from your home to the cable office). Such two way communication allows more sophisticated services like VoD to work — and that's why most cable companies began offering both digital cable and cable modems (which use the same upstream infrastructure) at around the same time.

Getting the benefits

So what does digital cable offer you (and your HDTV)? A lot actually (though, to repeat ourselves, HDTV isn't part of what's offered):

- **More channels:** Compressing each channel digitally allows more channels to be sent to your HDTV.

✔ **A high-quality NTSC picture:** Because channels are transmitted digitally, they are less prone to suffer from interference and "noise" than analog cable.

✔ **High-quality audio:** Stereo audio is provided with most programs; some channels may include a Dolby digital surround-sound audio stream.

✔ **An on-screen program guide:** Most cable systems include an interactive program guide that lets you view a week or so's worth of shows, view details about each program, and set up reminders and timers for recording.

✔ **Access to Pay per view:** Unlike old fashioned pay per view (PPV) programming, you don't need to call the cable company to watch a show (like a movie or sporting event) — you can simply press a button on your remote to order.

✔ **Access to VoD:** The flagship service in any digital cable package is video on demand. This service uses the two-way network and on-screen guide, and allows you to select any movie or program from a library of programs retained on a server in the cable company's network — when you hit Play on your remote, the program begins streaming down to your set-top box. You can pause, fast-forward, and rewind, just as you can with a DVD.

Getting the digital box

Today (though this *will* change very soon), you need a digital-cable set-top box in order to take advantage of all these neat digital-cable services. If you have an HDTV set-top box from your cable company, you probably already have a box that can take advantage of digital cable. (There are usually only a few HDTV channels on any cable system; it makes no sense to build a box for HDTV alone.)

Like the HDTV-specific set-top boxes, digital cable set-top boxes are rented or leased to you (the customer) in a variety of different ways — such as a monthly box lease, a monthly service fee, or a combination of those. Look for set-top boxes that have extra features — for example, Pat is currently using a set-top box from Scientific Atlanta that supports HDTV and digital cable, *and* has a built-in DVR (digital video recorder — see Chapter 12 for more info).

Analog Cable

The easiest cable signals to tune into are analog cable channels. Depending upon your cable system, all, none, or some of your cable channels will be analog (the latter case occurs when the first 50 or so channels on your system are analog, and all higher channels are digital).

Analog cable is nothing more (or less) than NTSC, analog, standard-definition TV programming sent over cables instead of the airwaves. With the vast majority of HDTVs (that is, any HDTV that is not just a monitor, and has an NTSC tuner built-in), you can view analog-cable programming by simply plugging your cable feed (the RG6 cable and F-connector — see Chapter 3 for more details on these cables) into the back of your TV.

If you're receiving digital cable or HDTV from your cable provider and are using a set-top box, you don't need to do anything special to tune into the analog channels. Your set-top box will automatically tune them in and send them to your TV.

Some plasma TVs (Chapter 23) and most front-projection TV systems (Chapter 22) don't have a built-in NTSC tuner. If you're not using a set-top box, you can always use the NTSC tuner built into your VCR (VHS, S-VHS, and D-VHS VCRs all have NTSC tuners built-in) as your NTSC tuner.

Going Boxless

Like many other electronic devices and services, digital cable (and HDTV cable) is going plug-and-play. With the help of the government and industry organizations (such as the Consumer Electronics Association), cable TV providers have gotten together with technology vendors (TV manufacturers and the folks who build the infrastructure components of a cable network) to come up with a system that lets customers kick the cable box to the curb!

As we write this, the first iterations of this "boxless" digital-cable system are beginning to be announced and trickle onto the market. We haven't been able to test them out. We're sure they work, and work well, but sometimes such transitions take longer than expected — so don't be surprised if it's a few years before this becomes commonplace.

So is the set-top box dead?

As the CableCARD system gains widespread acceptance, the need for the set-top box will decrease. After all, who wants extra devices plugged into their HDTVs (and sitting on the equipment rack) if they don't need 'em?

Understandably, cable companies (and the set-top box manufacturers who support them) don't see it quite that way. While CableCARDs will be offered (and will be required by the FCC to be an option for customers), the set-top box probably won't disappear completely.

Instead, the set-top box will become even more powerful — adding onto today's features like DVR with new features — such as home networking and whole-home DVRs that can record 6 or more channels and send the recorded programs to any TV in the home.

So what's going on? Two main things are happening:

- TVs are being built as *digital-cable-ready* (DCR) — just as regular analog TVs have been built as cable-ready for analog cable for years. These TVs have a QAM tuner built-in, and also have a slot for a "smart card" called the CableCARD.

- Cable companies are getting ready to rent these CableCARDs to their customers. A customer with a CableCARD and a DCR HDTV can simply plug the CableCARD into the HDTV — the card basically gives your TV "permission" to unscramble encrypted premium channels.

This permission-giving process is called *conditional access* in the cable business. No, we won't test you on that. Just thought we'd share.

With this DCR HDTV and the CableCARD, you simply plug the cable into the back of your TV and you're ready to go.

The first generation of CableCARD devices will be one-way only — so they will let you watch premium channels, but won't be able to communicate back "upstream" to the cable company. This means that the initial CableCARD deployments won't support digital cable services such as program guides, pay per view, or VoD. Eventually, an updated, two-way CableCARD will become available, allowing DCR HDTVs to access more of these advanced digital-cable features.

Cable companies hate paying for expensive set-top boxes, but they also like the concept of having a piece of gear in your home that controls much of your TV viewing (and maybe even your Internet surfing and phone calls, too!) — so don't be surprised to see your cable company still pushing set-top boxes your way, even long after CableCARD hits the streets.

Chapter 10

Rocket Science

● ●

In This Chapter

▶ Figuring out the pieces and parts

▶ Sifting through the offers

● ●

*W*ell, okay, rockets *did* have something to do with putting all those communication satellites up there. But that's pretty much where the science ends and the good news begins. Satellite service providers are not only a great alternative to cable for regular TV programming, they are also a primo source for HDTV. The satellite companies were among the first to offer HDTV — using HDTV as a serious competitive advantage to counteract cable companies' digital cable offerings, such as video-on-demand (VoD).

With its very small (usually 18- to 24-inch) dishes, *Direct Broadcast Satellite (DBS)* allows all but the most remote viewer to receive HDTV economically. Even author Danny on his island two miles out at sea.

Planning a Satellite TV System

Getting satellite TV for your HDTV isn't all that difficult, but it is a bit more complicated than cable TV (where most of the sophisticated equipment belongs to the cable company, so they handle all the maintenance and installation).

The first step towards satellite involves picking a provider (imagine that, a choice instead of a monopoly!). Once you've picked a provider, then you can choose the appropriate equipment — a satellite dish and a receiver.

Service

You must subscribe to a provider if you want HDTV channels from a satellite. Three TV service providers have significant HDTV programming:

- ✔ DIRECTV (www.directv.com)
- ✔ Dish Networks (www.dishnetwork.com), owned by EchoStar
- ✔ Voom (www.voom.com), owned by Cablevision

The last half of this chapter guides you to select a provider for satellite HDTV.

Equipment

For HDTV (or any TV, for that matter) by satellite, you need equipment to use the satellite signal: a receiver, an antenna, a TV, and cables to connect them all.

Want more information than that? Can do!

Receiver

Satellite HDTV requires an HDTV *receiver* (sometimes called a *tuner* or *set-top box*) that works with your DBS provider.

If you already have a DIRECTV or Dish Network receiver, it probably *doesn't* receive HDTV. Check the Web sites for DIRECTV (www.directv.com) or Dish Network (www.dishnetwork.com) for the latest HDTV receivers.

Except for special HDTV connections, an HDTV satellite receiver usually looks and works just like a typical satellite TV receiver. Most HDTV satellite receivers include two essential items:

- ✔ A receiver for satellite-based HDTV channels
- ✔ A *terrestrial* broadcast HDTV tuner so you can pick up local HDTV channels that aren't on your satellite lineup

Local HDTV broadcasts are transmitted *digitally*. Expect any HDTV satellite system to get a fine picture from local HDTV broadcast stations unless the local signal has to travel so far that it's just too weak when you get it. Chapter 8 gives you the lowdown on receiving over-the-air broadcasts of local HDTV channels.

Why might I need a new dish?

HDTV signals are transmitted in their own *orbital slots* (from different satellite locations, in other words). For example, Dish Network broadcasts HDTV from the 110-degree slot, which many older Dish Network dishes (that's a mouthful!) can't tune in. In those cases, you'll need a new satellite dish that incorporates low-noise blockers (LNBs) that can tune in these new slots before you can receive the HDTV signal.

A few TV sets have *integrated* DBS receivers. Each receiver is for only *one* DBS provider, so you need to decide on a provider before buying an HDTV set with an integrated receiver.

If you want a *digital video recorder (DVR)* in your HDTV satellite system, it should be part of the satellite receiver. Later in this chapter, we tell you about DVRs that are available for DIRECTV and Dish Network. Chapter 12 has the whole DVR story.

Antennas

Even if you have a DBS dish, you may need a new dish for HDTV. DBS providers have launched new satellites to transmit HDTV (and local channels and other services), because "normal" DBS satellites are pretty much full to the brim with regular TV programming.

The DBS company usually installs *two* HDTV antennas for you:

✔ A small broadcast antenna that receives local HDTV channels

✔ The special HDTV dish for your satellite channels

If you already have DBS service, it's easy to check whether you need a *new* satellite dish to get HDTV satellite channels from the *same* DBS provider:

• If you have DIRECTV service, look for a DIRECTV logo on Channel 99. If you see the logo, you're set for HDTV from DIRECTV. Otherwise you'll need to install a new dish (or in a few cases, purchase an upgrade kit for your existing dish).

• With Dish Network, try tuning to channel 9900. If you receive this channel, you've got the right dish — otherwise you'll need to install a new dish.

If you need a new satellite dish, let a pro install it. Running cables and aiming the dish can be pains in the hind end. There's no real advantage of doing it yourself, especially if you can get *free* installation in a package when you subscribe.

HDTV-ready TV

Your new HDTV-ready TV connects to the satellite HDTV receiver the same way it connects to a similar cable box. Chapter 1 and Chapter 2 guide you to the right HDTV-ready TV for your home.

If you don't have an HDTV yet, you can use an HDTV satellite receiver with your existing TV. You won't see HDTV quality pictures with such a set-up, but it works. You might do something like this, for example, if you're waiting for a specific HDTV, but you're ready to install or upgrade your satellite TV system now.

Cabling

An HDTV satellite system needs three kinds of connections:

- ✔ **Satellite-dish-to-satellite-receiver:** When the signal arrives from outer space, it has to get to the receiver before you can see it. No rocket science there.

    If you already have a DBS system, you probably can use the old cable to connect your new HDTV dish to your new HDTV satellite receiver. But let a pro install and aim your dish.

- ✔ **Local-antenna-to-satellite-receiver:** Chapter 8 guides you for receiving HDTV broadcast stations on HDTV devices.

- ✔ **Satellite-receiver-to-TV:** Chapter 3 covers basic video connections to an HDTV-ready TV.

Availability

Most U.S. homes can receive HDTV by satellite. Here's how to make sure *you* can receive it:

- ✔ The best way to find big problems or special needs is to check each provider's Web site. Zip-code-based information portals can tell you what's needed for your area.

- ✔ If you are in an apartment, condo, or a homeowner's association, check your local rules about dish placement. In the U.S., you must be allowed to install a dish (that's a federal regulation), but the dish must be located in a space that only you have access to (like a private patio) and not on common space (like the rooftop). If your permitted placement doesn't have a southwestern sky view, you will need to negotiate with your landlord or condo/homeowner's association about installing the dish elsewhere (for example, the roof). Check out the FCC's Web site (www.fcc.gov/mb/facts/otard.html) for the details on this ruling.

✔ If you are in Alaska or Hawaii, you might need a larger dish.

For example, DIRECTV advises its customers that subscribers in Anchorage or Fairbanks might have to install an 8-foot dish to get their signals. (You can probably get by with a 4-foot dish in Juneau.) If you live to the north and west of these locations signal strength continues to drop off, but we're guessing if you live that far north of the Lower 48, then you probably know this as a matter of course.

✔ Special *international* channels, such as Chinese- or Spanish-language channels, may come from other satellites. They require a *multi-satellite* dish and *visibility* to all those satellites.

Choosing a DBS Provider

Your home probably can receive all three satellite-HDTV providers. If you can decide by *features* and *price* (most folks who live in medium-sized and larger cities can decide that way), the trick becomes sorting through the offers *du jour*.

At the time this was written, Voom was offering a limited-time free-equipment-and-installation deal to drum up business, while Dish Network was packaging an HDTV, HDTV receiver, and its service for $1,000 installed — and was running a contest to give one lucky winner all the necessary equipment and a free year of Dish Network's HDTV service.

Special deals like the Voom and Dish Network offers discussed in the previous paragraph come and go, so you need to check with all three vendors for their packages when you are ready to buy.

Programming

All the HDTV satellite providers have the common channels you expect in any cable or satellite system. But actual HDTV channels vary from provider to provider.

Voom channels

Voom has the most HDTV channels, probably because it was launched for the explicit purpose of delivering HDTV via satellite. Voom had 39 HDTV channels at the time this was written (far more than any other service).

Voom has the usual HBO, Showtime, and ESPN channels. Some Voom programming is specifically for Voom's lineup. These include such unusual offerings as

- ✔ Equator HD, which broadcasts "intriguing and visually stunning sights and sounds that capture the world's most unique people and places" — that is, documentaries and other visually appealing fare

- ✔ A channel of beautiful HDTV images and music

- ✔ Special events, such as the June 2004 sale of Eric Clapton's guitar collection

At Voom, you can also watch Westerns on the Gunslinger channel, plus classic or epic movies (or epic classics, no doubt!).

Dish Network channels

Dish Network offers eight network HDTV channels (including HBO, Showtime, TNT, Discovery and ESPN), plus some special HDTV-only content:

- ✔ An HDTV Pay Per View channel, featuring recent movies and some IMAX films

- ✔ HDNet, which carries live sporting events, and a variety of dramas, documentaries, and other programming

- ✔ HDNet Movies, which shows (you guessed it!) movies from Warner Brothers, Sony Pictures Television, and independent sources

Voom or poof?

If you are unwilling to take a bit of a risk, Voom might not be the way to go — though Voom's current (as we write) offer of free hardware basically eliminates any risk to you, besides the inconvenience of switching to a different provider. In its first six months of operation, Voom lost a fifth of the customers who tried the service, according to Cablevision's SEC filings. The company has also decided to go with a bigger dish to enable subscribers to "see" a new satellite source that will provide capacity for more programming. The company is replacing its previous dishes for free, which is likely to increase its already-massive losses (just under $500 million this year) and depress its chances of succeeding as a spinoff from parent company Cablevision.

It might not hurt to take a leap of faith. Customers who get the service have raved about the HDTV menu. If Voom fails financially, the company's assets will likely be picked up by another company (maybe one of the other two satellite companies).

Internet and HDTV with one dish

It may be the industry's best-kept secret, but it's possible to get both high-speed Internet (via DIRECWAY) and DIRECTV's HDTV service from a single installed dish. The dual dish comes with *three* connecting cables:

✔ A cable to the HDTV satellite receiver.

✔ Input and output cables for Internet service. The input and output cables connect into a *high-speed modem,* which then connects into the computer.

You can share DIRECWAY Internet service with more than one computer in your home by either connecting computers to your DIRECWAY service through a home network or purchasing a DIRECWAY modem that can connect more than one computer.

DIRECWAY is a godsend for anyone living on an island or remote mountaintop, where you'd never get any other Internet access (ask Danny — he knows).

Compared to other Internet services, DIRECWAY has a couple of drawbacks:

✔ DIRECWAY is a *fast* Internet service, but it adds a *delay* in all your Internet activities. Its signals must travel 23,000 miles up to satellites and then the same distance back down. That adds a half-second to every round trip. The delay isn't significant for e-mail, text messaging, and Web-surfing, but you'd notice it if you play games or use voice over the Internet.

✔ DIRECTV and DIRECWAY haven't been offered together as a special deal. (Cable and telephone companies often give you a discount if they sell you TV and Internet service together.) But that can change. If you're considering both DIRECTV and DIRECWAY, look for combination deals that may have been created since this book was published.

If you intend to install DIRECWAY, ask for the *dual dish* at a DIRECTV retailer.

DIRECTV channels

DIRECTV offers eight channels of HDTV broadcasts of popular networks (including HBO, Showtime, TNT, Discovery, ESPN, and Spice — use your parental blocking if you need it!), plus a pay-per-view offering.

DIRECTV sells its HDTV for about $11 a month. The company sells its equipment through major retailers such as Best Buy and Circuit City and directs customers to the nearest reseller from its Web site.

Digital video recorder (DVR)

Digital video recorders let you do pause, do instant replays, and time-shift your favorite programming. (We cover DVRs in depth in Chapter 12.) A DVR is *more powerful* and *easier to use* with an HDTV satellite system if the DVR is *included* in the satellite receiver.

As of this writing, you have a couple of options if you want an HDTV satellite system with an integrated DVR. These options have the same basic capabilities:

TIP

- ✔ **DIRECTV:** Its HDTV receiver can come integrated with a *TiVo* digital video recorder.

  We prefer the TiVo user interface and overall experience.

- ✔ **Dish Network:** Its own DVR (the Dish Player-DVR 921, $999) doesn't use TiVo technology.

As of this writing, Voom doesn't offer a DVR, though the company has promised one before the end of 2004.

Part IV
Movie Machines

By Rich Tennant

@RICHTENNANT

"Harry says his new HDTV is so realistic, the remote has a 'CAST' button."

In this part . . .

*H*ave you noticed everything in HDTV boils down to acronyms? Well, in this part, we keep up that tradition, talking about DVDs, DVRs, and VCRs, and we do so ASAP PDQ!

DVDs are responsible for the boom in digital video content, just as CDs sparked the boom in digital audio in years past. The low cost, ubiquitous availability, and sheer massive storage space of DVDs revolutionize everything from blockbuster to direct-to-DVD movies — it's even spawned a new business of getting DVDs in the mail, NetFlix. So we spend some time looking at DVDs and DVD players in this part, mainly to bring you up to speed on all this great content you can use to drive your HDTV experience. We also talk about the future — great new DVD technology that will bring true HDTV programs to the DVD format.

We'll also talk about the latest and greatest thing to hit consumer electronics, the DVR (digital video recorder). Called TiVo by a lot of people, this product category has transformed how people interact with their TV set. The DVR lets users watch what they want, when they want, even if they missed it. Got that? You will. Combine this with the availability of full season's worth of popular shows on DVD, and you've totally transformed the way TV is watched.

Finally, the VCR is still alive and kicking. If you are like us, you've got shelves of VHS movies and older VCRs spread across the house. With new HD camcorders and HDTV content over the airwaves, you might want to upgrade to a new digital VHS VCR. We tell you the pros and cons of D-VHS options, and what's available on the market.

Chapter 11

DVDs

DVDs probably will be your main source of non-broadcast video content for HDTV viewing. DVDs have gone from zero to 1,000 miles per hour in no time. More than 70 percent of American households have adopted the technology in less than seven years.

The biggest questions about DVD don't have anything to do with today's DVDs and DVD players. You can buy a serviceable DVD player for $30 — if you don't mind getting trampled at Discount City — or a high-quality unit for $200. The confusion is about the *next* generations of DVDs — the profusion of recordable DVD formats, and the competing high-definition DVD formats.

With a few expensive and rare exceptions, you *can't* buy HDTV DVDs now, but some standard DVD players are *better* for HDTV.

In this chapter, we start with the basics of standard DVDs and DVD players, and guide you through the confusing territory of DVD recorders. We follow with all of the *form factors* (packages) that contain DVD players — DVD players are in all sorts of electronics gear. We finish by putting on our Crystal Ball Spectacles to see high-definition DVD formats that may be available soon.

Learning about the DVD disc

Digital Video Discs (sometimes they're called Digital *Versatile* Discs — this usage has pretty much disappeared) are simply *optical* (laser-readable) storage media — like a CD *(compact disc)*. Like the CD, the DVD is 12 cm in diameter, and consists of *layers* — the top layers protect the DVD, and the inner layers contain tiny pits that can be detected and "read" by a laser beam.

Today's DVDs can hold at least 4.7 gigabytes of data on one side (compared to about 800 megabytes on a CD). That's enough for about two hours of *standard definition* (NTSC) video.

Currently, there are two ways to put more video on a DVD:

- ✔ **Add a layer:** DVDs can be designed with a second layer of data — basically another layer of *pits* at a different depth within the DVD. The laser in a DVD player can focus on this layer and ignore the other layer entirely.

- ✔ **Use both sides:** DVD data can be on both sides of the disc.

  DVDs seldom use both sides of a disc to show different *parts* of a movie. Double-sided DVDs usually are seen when

  - A 16:9 *widescreen* version of a movie is on one side.
  - A 4:3 *pan-and-scan* version is on the other side.

DVDs can be *both* double-layered and double-sided — holding up to 18 gigabytes of data.

There are a couple of other ways to fit more data onto a DVD. We cover both of these at the end of the chapter:

- ✔ A new *blue* laser that can use smaller pits
- ✔ New data-compression systems

Dealing with today's DVDs

Today, DVDs usually are the best source for playing prerecorded movies on an HDTV. Here's why:

- ✔ **Distortion-free digital image:** It isn't HDTV, but DVDs offer a clear, colorful and sharp picture at higher resolution than either VHS or PVRs currently offer (with the exception of D-VHS VCRs, which are almost impossible to find).

- ✔ **Progressive scan:** With a progressive-scan DVD player, your progressive-scan HDTV can be fed a non-interlaced picture (often providing better quality than if your HDTV had to "de-interlace" the picture itself).

- ✔ **True widescreen:** With discs labeled *anamorphic* or *formatted for widescreen*, you can get a full widescreen 16 x 9 picture for your HDTV without techniques that reduce the resolution of the video, such as letterboxing.

There's a wide range of prices for good DVD players — from well under $100 up to the thousands for "high-end" models.

Essential features

We insist on all of these features in a DVD player for any HDTV system. You should be able to find an affordable DVD player with

- ✔ **Progressive scan** (described in Chapter 21)
- ✔ **3:2 pulldown** (described in Chapter 21)

   If your monitor has a good 3:2-pulldown system, you can get by without it in the DVD player, but you aren't likely to find a good DVD player without built-in 3:2 pulldown.

- ✔ **Component video outputs** (described in Chapter 3)

Useful features

Depending on how you use your DVD player, these features can make your HDTV system more useful and convenient:

- ✔ **Built-in surround-sound decoder:** If your old surround-sound system doesn't have a built-in decoder for current formats, this can add Dolby Digital and DTS (which we cover in Chapter 18).
- ✔ **Extra audio formats:** All DVD players also play *audio CDs,* but some go beyond the call of duty with these audio formats:
   - SACD
   - DVD-Audio
   - MP3 audio files on homemade CDs
- ✔ **Multidisc changer:** A multidisc changer is handy if you watch lots of movies (or use it as a CD player for a party).

   Single disc players are often more *reliable* than multi-disc changers.

- ✔ **DVI-D output:** If your HDTV has an otherwise unused DVI-D input, you may consider this feature. Some DVD players with DVI-D outputs feature an internal *scaler* (described in Chapter 6) to better match the DVD output to the HDTV. DVI-D is described in Chapter 3.

Deciphering DVD Recorders

If you could travel through time back to (say) 1986, you could really blow the minds of anyone you met by saying that less than 20 years later, you could make your own CDs at home. It's like saying you can make your own jet airliner in the garage. But you can't even buy a PC these days without a CD burner built-in.

In less than seven years on the market, the DVD has also become something you can make yourself:

> ✔ Most DVD burners are in personal computers.
>
> Extra capacity makes the DVD format really useful in general with PCs. As quickly as computer hard drives have grown, an 800-megabyte CD now seems puny.

> ✔ You can buy a *standalone* DVD recorder. It's an A/V component that can also be your DVD player.

Recordable DVDs labeled with an "R" can be written *(recorded)* once. "RW" discs can be rewritten *(changed)* thousands of times.

A DVD recorder *can't* copy most commercial DVDs, which are copy protected. Your DVD recorder can detect signals in the video stream and prevent copying. You can record TV programs, most VHS tapes, and homemade content (such as camcorder tapes).

The biggest decision to make when considering a DVD recorder is the *format.* The consumer electronics industry (and PC industry) doesn't agree on a standard format for recordable DVDs. There are three competing DVD recording formats (which aren't always compatible with each other):

> ✔ **DVD-R/RW** ("DVD dash") discs are the most compatible with other DVD players, so a DVD-R or DVD-RW disc is most likely to play on your mom's old DVD player.
>
> ✔ **DVD+R/RW** ("DVD plus") discs are essentially as compatible as DVD- discs (some DVD+ vendors claim *more* compatibility — we think it's about a draw).
>
> DVD+ can record dual-layer discs with double capacity.

> ✔ **DVD-RAM** discs are mainly for computer data storage, and can be rewritten (like DVD-RW and DVD+RW discs).
>
> DVD-RAM is the least compatible format with other players.

Finding DVDs in Unusual Places

The most common way to buy a DVD player is to get your hands on a standalone DVD — but that isn't the only way to get DVD into your home. DVD players have been slotted into all sorts of different home entertainment gear — usually in an effort to save space and money by creating "all-in-one" devices that can feed different signals into an HDTV.

Xbox and PlayStation 2 game consoles can be DVD players, too!

Home theater in a box

The most common "all-in-one" DVD sources are the *Home Theater in a Box* (or HTIB) systems from just about every major A/V gear manufacturer. These systems usually include

- An *A/V receiver* (see Chapter 18) with a built-in DVD player
- A complete set of surround-sound speakers (including a subwoofer) for Dolby Digital sound from DVDs and HDTV

Figure 11-1 shows an HTIB system.

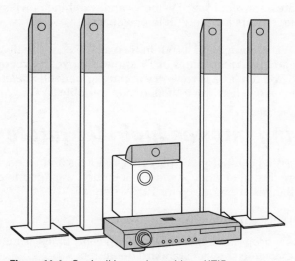

Figure 11-1: Get it all in one box with an HTIB system.

Chapter 19 covers sound systems for HDTV and home theaters.

DVD/VCR

Some manufacturers squeeze a DVD player and a VHS VCR deck into a single chassis.

These devices use one remote control and one power supply, but usually *two* sets of video outputs that connect to your HDTV:

- ✔ Component video for DVDs (plus S-video and composite)

  Avoid DVD/VCR units that don't have component video for the DVD player. You really want be able to use component video to connect the DVD player to your HDTV!

- ✔ Composite video for VHS (or S-Video if the VCR portion supports that system)

You can't record commercial DVDs onto the VCR in these units, because of a copy protection system in DVDs.

DVD/PVR

The neatest development we've seen recently is the integration of DVD players (and particularly DVD *recorders*) and PVR (personal video recorders — Chapter 12 covers these devices). It's the latest and greatest for recording TV shows and watching movies — the two replacements for the VCR in one chassis!

With a DVD recorder and a PVR in the same box, it's really easy to make archived recordings of TV shows that you've recorded on the PVR. You don't have to worry about permanently *deleting* programs if you can just *move* them to a recordable DVD.

Peering into the high-def future

Today's DVDs can't store or play HDTV movies and video, because they were designed when standard definition was "good enough." Three changes are required for HDTV from a DVD:

- ✔ DVD discs need more *storage space.*

  HDTV programming can have up to eight times as much data as the same show in standard definition.

- ✔ DVD players must *read* these new DVD discs.

- ✔ DVD players must *output* HDTV signals.

Fortunately for HDTV addicts, the major players in the DVD industry are working furiously to develop high-def systems. Unfortunately, these companies have taken separate (and incompatible) paths to DVD HDTV nirvana.

With these competing systems, we expect another *VHS versus Betamax* battle in the market. One format probably will end up "winning" — if you pick the wrong system, you'll have obsolete equipment much sooner than you expected. Don't be the first on your block to buy one of these machines (unless you're loaded and don't mind losing the investment). Give the market some time to sort out. We hope that eventually universal players will play both types of discs.

The battle will be over both *technologies* and *standards*.

You need a *new DVD player* for the following HDTV formats. However, the promoters of these standards promise their new DVD players also will be compatible with your *old* DVDs.

Blue-laser systems

One way to increase DVD capacity is to switch to *blue-laser* technology. Blue lasers can read smaller pits in the DVD disc than red lasers (that's because a blue laser has a shorter *wavelength* than a red laser). Smaller pits mean *more* pits (and data) fit on a disc.

There are a couple of competing blue-laser standards.

Blu-ray

The largest group of companies supports a blue-laser system called *Blu-ray.* The Blu-ray bandwagon includes Sony, Panasonic, RCA, Pioneer, Philips, Samsung, and LG.

A single-sided Blu-ray disc can hold up to 25 gigabytes of data on a single layer (50 gigabytes on a dual-layer disc), which is more than enough to handle most movies in HDTV resolution.

Blu-ray players aren't widely available as we write. At least one (very expensive) model is available in Japan.

HD-DVD

High-Definition DVD (HD-DVD) is a blue-laser system supported by Toshiba, NEC, and the DVD Forum (an industry group that promoted the original DVD format). HD-DVD systems use a blue laser with larger pits in the DVD disc than Blu-ray.

Larger pits should make HD-DVD discs easier to manufacture, but they hold less data than Blu-ray discs. To make up for it, the HD-DVD folks use more aggressive *video compression* technologies (like Microsoft Windows Media) so longer movies fit on a disc.

Compression basically shrinks the amount of storage space needed by a video by discarding "unnecessary" bits of the video. This can cause some degradation in picture quality — how much you lose is subjective, and it depends on which compression system is used. A good compression system is nearly unnoticeable.

As we write, HD-DVD players aren't available from any manufacturers — they're still in prototype stage.

Compressed red-laser systems

There are a couple of efforts to use *video-compression* technology to fit HDTV signals onto regular *red-laser* DVDs. The discs would look like standard DVDs, but a new DVD player would be required to *decode* these compression methods.

Compressed red-laser systems can make HDTV DVD movies in existing factories, without new equipment. It doesn't make much difference to you, unless using the same factories lowers what *you* pay for HDTV DVDs.

The benefits are for the big studios making the DVDs themselves:

- ✔ Time Warner supports a red-laser system called *HD-DVD9*, which uses Microsoft Windows Media compression system.

- ✔ A group of manufacturers in China is proposing the *EVD* compressed system.

Chapter 12

Getting Into DVRs

*B*esides HDTV itself, we think that the rise of the *DVR* (Digital Video Recorder — some folks call them PVRs, or Personal Video Recorders, too) is the single biggest thing to happen to TV in our lifetimes. Yeah, we know that sounds like an exaggeration, but trust us — using a DVR is almost a life-changing experience. Once you start using a DVR, you'll never (never ever) think about TV the same way again!

In this chapter, we give you a good dose of DVR background and information, then cover the (few) DVRs that can record HDTV signals. We also discuss DVRs more generally — after all, most features and "goodies" found on today's standard-definition DVRs will eventually appear again in high-def versions of the same units. Finally, we cover some interesting DVR variants (such as the DVR/DVD recorder combo that Pat finds so compelling).

 Most DVRs in this chapter (and most DVRs you can buy today) *can't* record or play HDTV content. A very small number of HDTV-capable DVRs are available, but we expect that very soon HDTV DVRs will be common.

DVR 101

A DVR is, at the most basic level, a digital replacement for your VCR. Just like a VCR, a DVR can record TV programs — instead of using a tape for this purpose, a DVR records TV digitally, on a computer hard drive inside the DVR.

Learning the benefits

If the only benefit of a DVR were the replacement of the tape with a hard drive, we'd be sold on it. But as they say in the Ginsu knife commercials —"Wait, there's more!" DVRs can also

- **Pause live TV:** Usually, DVRs automatically record about an hour's worth of the show you're currently watching. So you can hit the pause button, run to the bathroom or kitchen or answer the phone, and pick up later where you left off.

- **Rewind live TV:** You can rewind a scene that you missed. This was done a lot during a certain halftime show (with a certain "wardrobe malfunction").

- **Record a show you're watching with one touch:** If you must leave the room suddenly, just press a single button on the remote to save the whole show (including the part you've already watched).

- **Record an entire series:** You don't have to set up separate recordings for each episode of a series. Just select one episode and turn on series recording. Your DVR automatically finds every episode of that show on the programming guide and records it for you.

All DVRs have a few common elements:

- The hard drive for storing video

- A *GUI* (graphical user interface) that lets you control the DVR with your remote control

- An on-screen programming guide for scheduling recordings (no more need to manually enter dates and times!)

- A telephone or Internet connection for updating the programming guide

The best-known DVRs come from a company called TiVo (www. tivo.com) and its partners. In fact, the name TiVo is widely used to refer to DVRs in general — it's even used as a verb as in, "I didn't watch it yet, but I TiVo'd *Deadwood.*"

Making the connection

Connecting a DVR to your HDTV system is simple. You want your DVR to be *inline with* (connected between) your TV source (antenna, cable or satellite) and the HDTV itself.

✔ This is incredibly easy if the DVR is part of the TV source (either a satellite receiver or cable set-top box).

✔ If you have a *standalone* DVR, you connect it to the standard-definition outputs of your satellite or cable set-top box (or to your antenna cable) before those signals reach your HDTV.

Today, there aren't any HDTV standalone DVRs on the market — so you have only standard-definition signals on your DVR. That's because the common self-contained DVRs that are most popular today have not yet been converted to handle HDTV — manufacturers aren't yet convinced that there's a big enough market for HDTV viewers using over-the-air antennas.

You could use one of the DBS satellite HD DVRs we discuss in the section of the chapter that immediately follows to record over-the-air HDTV signals. In this case, the DVR is also your HDTV tuner. If you're not subscribing to a DBS service, this is a really expensive ($1,000) way of getting an HDTV-capable DVR just for broadcast channels.

Figure 12-1 shows a typical DVR set-up with a satellite or cable set-top box and a VCR.

HD DVRs

If you own or plan on owning an HDTV (and we assume you do, since you're reading *HDTV For Dummies*), your ultimate DVR records and displays HDTV programs in their native HDTV resolutions (720p or 1080i). After all, your favorite HDTV shows deserve the deluxe DVR treatment as much as your standard-definition favorites.

As we write, there are few HDTV-capable DVRs on the market. Making a high-def DVR is a cost-engineering challenge, not a technical challenge. Today, a high-def DVR isn't easy to build at prices consumers will pay, because a high-definition DVR requires

✔ A much bigger hard drive

✔ A more powerful graphics chip

✔ More expensive connectors (like DVI-D or component video)

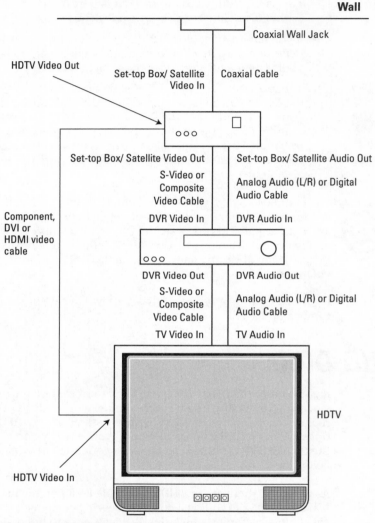

Figure 12-1: Squeezing a DVR into the picture.

 The prices of high-def DVR components are coming down rapidly. We expect that HDTV DVRs soon will be common and affordable.

If you want an HDTV DVR today, you have a few choices:

  ✔ **From your cable company:** A few cable companies offer digital cable/HDTV set-top boxes that include HDTV DVR features.

One example of this is Scientific Atlanta's Explorer 8000HD, which you can look into at the following site:

`www.sciatl.com/products/consumers/Exp8000HD.htm`

You can't buy one of these for yourself, but if you're lucky, your cable company will rent you one for a monthly fee. The Explorer 8000HD (shown in Figure 12-2) contains a 160 gigabyte hard drive that can record up to 20 hours of HDTV programming.

Not all cable-company "HDTV" DVRs actually record in HDTV — some combine an HDTV tuner with a *standard* DVR. This DVR is better than nothing, but know what you're getting for your money.

✔ **In your satellite receiver:** Each of the major satellite TV providers have announced deals to provide their customers with set-top boxes that can record HDTV — for the low, low price of about $1,000. We talk about the satellite service providers (DIRECTV, Dish Network, and Voom) in Chapter 10.

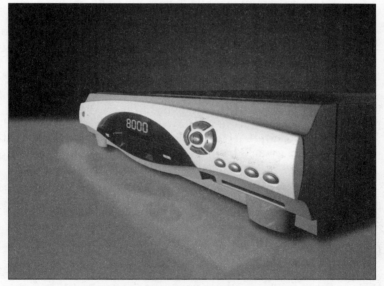

Figure 12-2: The Scientific Atlanta Explorer 8000HD can record HDTV.

Standard DVRs

If you are one of the lucky few with the budget, the right HDTV service provider, and the right alignment of the stars, you might be able to get an HDTV-capable DVR *right now.* For the rest of us, a standard-definition DVR is the only option until HDTV DVRs become more widely available (and affordable!).

What to look for

DVRs are made by many manufacturers, and are available at a wide range of prices. Here are the key things we look for when deciding which DVR is right for us (these criteria can apply to an HDTV DVR, too):

- **Purchase or rent:** Your main question probably is, "How much is this thing going to cost me?" The answer depends upon where the DVR comes from. Some units are offered to consumers for outright purchase; others are rented from a cable company or other provider on a monthly basis.

Do the math on this — rental is initially cheaper, but may be more expensive if you keep the DVR for a long time.

- **Programming guide fees:** Many DVRs require a monthly service fee to access the on-screen program guide (it makes the DVR work, and you should consider it mandatory). Some, like those rented from a cable company, may not charge for this, or include this cost in the equipment rental fee. In some cases, you can save money by purchasing a "lifetime" subscription.

Lifetime subscriptions only are a good deal if the company that sells them stays in business for the lifetime of your DVR. DVR manufacturers have gone *bankrupt* — folks with lifetime subscriptions weren't affected, but for a while it wasn't clear what would happen. If your DVR vendor goes bust, you and your subscription could be out in the cold.

- **Hard-drive size:** The main technical specification to check in a DVR is hard drive size — often described as *hours of programming* that fit on the hard drive.

Hours can trick you — they depend on video *quality.* Compare the same video quality when comparing time.

Check whether the PVR allows for adding external hard drives to add capacity in the future. Look for a DVR that includes either *FireWire* (also called *1394*) or *USB 2.0* connectors for this purpose. Read the manual (the fine print!) — some DVRs have these connectors, but they aren't "turned on," so you can't use them for this purpose.

- **Number of tuners:** We recommend a two-tuner DVR, so you can either record two programs at once or watch one while recording another.

Some DVRs only have a single TV tuner built-in. You can only record one program at a time — if this single-tuner DVR is built into your cable or satellite set-top box, you might be able to watch only the program you're recording.

✓ **Hardware features:** The most basic DVRs simply have connections for TVs and TV source signals. Others can connect to home networks (via Ethernet) or even wireless LANs for sharing content between DVRs within the home.

We recommend Ethernet connections so you can share!

✓ **Software features:** Some DVRs (like those made by TiVo or ReplayTV — www.replaytv.com) are built around very sophisticated GUIs that can do things like recommend shows you'd like (based on your viewing habits), display pictures or play music that's stored on your PCs, or even incorporate Internet content. Others (for example, some of the DVRs incorporated into cable set-top boxes) are more basic.

If you want all the fancy interface stuff, you're probably best off with either a *TiVo* or *ReplayTV-powered* DVR.

Finding a DVR that fits

When buying a DVR, you need to consider how it fits into the space you want to put it in — its *form factor*. Some DVRs are self-contained units — in other words, a chassis with a DVR inside, and nothing else. Others are incorporated into other devices, providing a multipurpose device that takes up less room on that equipment rack under your HDTV.

Among your choices are the following:

✓ **Standalone DVRs:** The most popular DVR is a standalone unit, like the traditional TiVo and ReplayTV models.

These DVRs are best suited for connecting to over-the-air *broadcast TV* or *analog cable* systems, though they can be used with a digital cable set-top box or satellite receiver.

✓ **Cable and satellite DVRs:** If you connect your HDTV primarily to digital cable or satellite service, you should consider a DVR that is built into your receiver or set-top box. These DVRs save space and receive the *digital* signals used by these services (which standalone DVRs can't do). Note:

- Satellite users usually *buy* the DVR/receiver.

- Cable users usually *rent* equipment from the cable company.

✔ **DVR recorder DVRs:** Our favorite standalone DVRs have a built-in DVD recorder. A DVD recorder solves the biggest issue most DVR owners run into — what to do with all those back episodes of your favorite show when your hard drive starts to get full. With a DVD recorder, you "burn" a DVD archive and clear your hard drive for new recordings.

✔ **Using your PC:** There are a couple of ways to use a PC:

- If you have a Windows XP Media Center PC, you *have* a DVR. These PCs already have the hardware and software to record TV programs; the stuff works just like a regular DVR.

- You can turn a standard PC into a DVR by adding some inexpensive hardware and software.

Chapter 16 covers the equipment you need if you want to use a PC as a DVR.

One way TiVo is fighting back is by turning to the Internet. TiVo has announced that it will offer a new service to its *broadband* (cable- or DSL-modem-equipped) customers — an Internet service will allow customers to download movies, video, and music directly to their TiVo DVRs. Basically, this service is designed to both supplement and to *bypass* regular TV sources — providing TiVo with a compelling service of their own to keep people buying TiVo DVRs (and to keep them subscribing to TiVo services). Check whether it's available at www.tivo.com.

Using your DVR on the Net

TiVo, one of the originators of the whole DVR concept, has been facing an uncertain future as DVRs are built into more and more devices (such as cable services' set-top boxes and Media Center PCs). Indeed the company faces a bit of a squeeze as cable and satellite companies develop their own PVRs that may be more convenient for cable and satellite customers than using a standalone TiVo.

Chapter 13

Taping Time

· ·

In This Chapter

▶ Viewing your old *Scooby Doo* VHS tapes on your HDTV

▶ Making the VCR to HDTV connection

▶ Understanding the advantages of digital VHS

▶ Buying your first (and the only) D-VHS VCR

· ·

*I*n our collective rush to sift through the $5 DVD bins at Wal-Mart, it's easy to overlook the good ol' VCR that sits precariously atop many of our TV sets at home. DVDs are cooler, packed with more data, and they're . . . well . . . digital.

But almost all of us have a lot of investment in the VCR — ranging from the scores of VHS files purchased over the years, to the recorded broadcast and home movies that make up our VHS collections. It may not be a big investment, dollarwise, but we often have an emotional investment in those old tapes! Even in the face of several generations of higher-quality video options — laser disc, DVDs, satellite, and now HDTV and DVRs — it's hard to part with our dependable VHS VCR.

That's because the venerable VCR is simply a useful device — and it has even been remaking itself for this new, high-definition world. In this chapter, we talk about how to get your old VCR hooked up to your HDTV so you can dust off those old taped episodes of *LA Law* (you '80s fan, you!); we also discuss some of the (few) new digital VCR models hitting the market that you might want to investigate to upgrade your taping capabilities.

Checking Out Your Digital Options

If you stroll into your neighborhood electronics store, or surf for VCRs online, you'll see three kinds of VCRs available today:

✔ **VHS VCRs:** Familiar to all, these are the standard VCRs you've seen in stores for almost 30 years. These VCRs use VHS video-cassettes and record a low-resolution TV signal — about 240 lines of resolution (a third of the 720 lines you get with HDTV!). These VHS VCRs can come cheap (less than $50) or expensive (more than $800); most include stereo audio capabilities (if so, they're labeled *HiFi*) and analog Dolby surround-sound capabilities.

✔ **S-VHS VCRs:** The S-VHS (or Super VHS) VCR provides a higher-quality, higher-resolution image — 400 lines, instead of about 240. In the midst of the craze for more resolution, you'd be forgiven for guessing that film buffs would have jumped at the first S-VHS VCRs back when they were introduced, but they've been slow to take off, for several reasons:

- With the early models, you had to buy more expensive (and harder-to-find), specialized tapes to record in S-VHS mode.

- You can't play an S-VHS tape in a regular VCR (but you can do the reverse).

- S-VHS VCRs themselves have been significantly pricier than the standard-format VCRs on the market.

- There were very few prerecorded movies available in the S-VHS format.

Lately, however, S-VHS VCRs have made a bit of a comeback, driven by lower equipment pricing — S-VHS VCRs start off at around $120 — and a new technology boost in the form of S-VHS-ET (Expansion Technology), introduced by JVC. S-VHS-ET allows an S-VHS VCR to record a full S-VHS signal on regular VHS tape. There's also the fact that higher-resolution source material from camcorders, satellite and digital cable transmission — and the ensuing display of recorded tapes on higher-definition displays — have raised the bar for video recorders as well. (You really notice poor resolution when it's 50 inches and 1280 x 720 in front of you!)

There's very little difference in pricing between a high-quality VHS VCR and an S-VHS VCR (really cheap VHS VCRs *are* a lot cheaper, of course). If you're already spending 100 bucks or so for a hi-fi VCR, we think you should find the extra $20 to get into an S-VHS model.

You'll also find a few VCRs on the market that are labled "Quasi-SVHS". These units *will* play back S-VHS tapes, but only at the lower VHS resolution. They're handy if you don't care about the higher resolution, but have some S-VHS (or S-VHS-C camcorder) tapes that you want to watch.

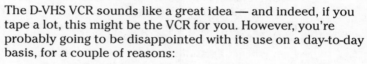

 D-VHS VCRs: If you're going whole-hog high-def, then you want to look at a high-definition VCR — specifically, the D-VHS VCR. D-VHS VCRs can play and record standard VHS and S-VHS formats. More importantly D-VHS VCRs can also record and play back in all the HDTV formats (discussed in Chapter 21), including 1080i, 720p, 480p, and 480i.

To record and play back HDTV, the D-VHS system needs an FireWire (also called i.LINK or IEEE 1394) connector to connect to an HDTV TV or a standalone HDTV tuner. Many HDTVs and HDTV tuners don't have a FireWire connection, however, because broadcasters and movie studios don't want you to be able to record HDTV programs. So be sure of your connection options before plunking down the cash for a D-VHS VCR.

The D-VHS VCR sounds like a great idea — and indeed, if you tape a lot, this might be the VCR for you. However, you're probably going to be disappointed with its use on a day-to-day basis, for a couple of reasons:

- D-VHS would be pretty cool, if many people weren't restricted from using it for its intended purpose — recording HDTV shows for later playback. Unfortunately, there's a fundamental disconnect between the HDTV connection provided on D-VHS decks (FireWire), and that found on the majority of HDTV tuners/set-top boxes (DVI).

- There are only limited number of movies available on prerecorded D-VHS videocassettes — truly a small number (less than a hundred) compared to what you can get on DVD (tens of thousands).

Many people who used to use VCRs a lot are focusing on DVRs (see Chapter 12) or are using DVDs, particularly as HDTV-capable DVD formats are being announced and heading to market (see Chapter 11). As a result, D-VHS VCRs have not really taken the market by storm — and really dropped in price recently as a result (they can be had for about $600). The D-VHS VCR category is largely heading for Laserdisc status (that is, dead), or at least confined to a niche market of D-VHS enthusiasts.

What will really nail the coffin of the D-VHS system is the advent of HDTV-capable DVD systems — particularly when these offer recording capability to users. Combine a high-def DVR with a recordable Blu-ray DVD system and you have the best of all worlds. No tapes to mess with, just pure, digital, skip-right-to-where-you-want-to-be-without-rewinding bliss.

Given that this is a book on HDTV and not on VCRs (and that the main high-definition angle to VCRs is D-VHS), we're going to spend the rest of this chapter talking about how to hook up your VCR into your HDTV system for best viewing. We also look at one of the only D-VHS products actually on the market as we write. If you want to know more about VCRs in general, try checking out `http://hometheater.about.com`.

Connecting VCRs to HDTVs

You often have a choice of connections/cable types to use when connecting source devices (such as VCRs) to your HDTV (Chapter 3 covers the details). Choosing which of these connections to use can be (initially) a bit confusing, but they fall into a hierarchy — and when you get a handle on that, matters get easier: Just pick the best connection option available on both ends of the connection.

When it comes to VCR connection options, we can make it even easier for you. You have three choices, depending on what kind of VCR you own:

- **Connecting VHS VCRs:** If you have a VHS VCR (not an S-VHS model), you have only two choices — you can connect using a composite video cable (and a pair of audio cables), or you can use the coaxial cable output of the VCR. We recommend that you use the composite video cable — it delivers a much better picture than coaxial cable.

- **Connecting S-VHS VCRs:** These VCRs add an *S-video* connector (that's where the name came from originally — even though S-video connections are now found on DVD players, set-top boxes, game consoles, and other source devices). S-video provides a *much* better picture than composite video, so use this connection (and a pair of audio cables) to connect your S-VHS videocassette recorder.

- **Connecting a D-VHS VCR:** You need to use *component* video cables to get HDTV signals into an HDTV display. So it should come as no surprise that the preferred connection method for the HDTV-capable D-VHS VCR is a triumvirate (yep, there's a troika — three — of them) of component video cables. You'll also want to use a digital audio cable (optical or coaxial) to take advantage of the digital surround-sound signals recorded on D-VHS high-definition recordings.

If you don't use the component video cable, you *can't* view HDTV recordings in high definition.

In addition to these "playback" connections, which enable you to view videocassettes played on your VCR, you'll also want to make a "recording" connection between your TV source and your VCR. Your options are many, and depend greatly on a whole host of variables regarding what else you've connected to your HDTV (and how it's connected). Here are some general rules to follow:

✔ If you have a D-VHS deck, and you want to record in HDTV, the answer is simple. You need to connect your HDTV tuner to the D-VHS using the FireWire (or iLINK) cable. There is no other option — which severely limits the usefulness of the D-VHS, as few HDTV tuners have functioning FireWire connections that will make this work.

✔ For a non-HDTV VCR recording, your options depend on two things: whether you've connected a home-theater receiver as part of your overall HDTV home-theater system, and whether you require a set-top box or satellite receiver for non-DTV TV channels.

  • If you're using a home-theater receiver (discussed in Chapter 19), take advantage of this device's *video-switching* capabilities. You can connect all your video source devices (such as DVD players, cable set-top boxes, and satellite receivers) into a set of *inputs* on the receiver. Then you can use the *outputs* on the back of the receiver to connect both your HDTV (for viewing) and your VCR (for recording). This is the best, and most flexible, way of connecting your VCR for recording.

  • If you don't have a home-theater receiver, you have to connect your VCR *inline* between your TV source and your HDTV. This could be as simple as plugging an antenna cable into the back of your VCR (to the *TV In* connection) and then using another antenna cable (coaxial cable) to connect the *TV Out* jack on your VCR to the Antenna input on your VCR. If you have a set-top box or satellite receiver, you need an extra set of S-video or composite-video cables to make this connection. Figure 13-1 shows a VCR installed inline.

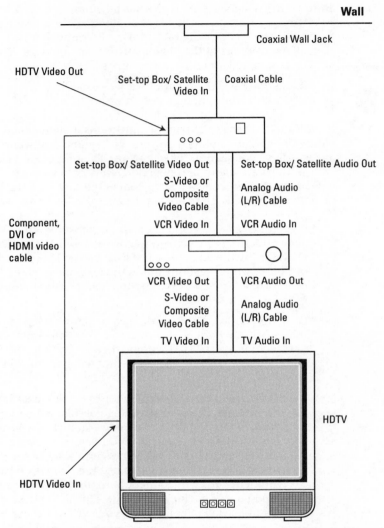

Figure 13-1: Connecting a VCR inline between a TV source and your HDTV.

Looking Closer at D-VHS

So if you have high-definition leanings (and why else would you be reading *HDTV For Dummies*?), you at least owe it to yourself to consider whether a D-VHS will work for you. So . . .

✔ Here's what you can expect to do with it:

- Record over-the-air broadcasts
- Watch D-VHS prerecorded tapes
- Store HDTV camcorder content to D-VHS

✔ Here's what you may not be able to do with it:

- Record cable or satellite broadcasts (many set-top boxes lack a FireWire port to send digital signals to the D-VHS, due to copyright concerns — and some that do have the capability *don't* have it *enabled* by software, so it's a useless connection that won't work).

✔ Here's what you can't do with it:

- Make tapes of your favorite Hollywood DVD movies. (A copy-protection system called Macrovision keeps you from doing so, and it's built into all DVDs and all VCRs. Even if you want to copy your favorite movie from DVD to VHS to play back on a VCR in your vacation home or somewhere else, you're just plain out of luck.)

If these benefits and limitations seem reasonable to you, then consider getting a D-VHS VCR. When it comes to playing and recording high-def signals, you have to go D-VHS or you're simply not getting all you can out of your HDTV lifestyle! No other format can match those high-res images, baby . . .

. . . at least not yet. When high-def DVD discs and players hit the market (and they've been announced as we write, but can't be bought in stores yet), their resolutions will equal those of D-VHS.

The D-VHS VCRs use high-capacity D-VHS tapes to record HDTV broadcasts at their full resolution. They can upconvert or down-convert HD recordings on playback to match your TV's display rate (even when connected to a regular analog TV).

D-VHS VCRs typically can do some other hand things:

✔ Record/play Digital VHS, Super VHS, Super VHS ET, and standard VHS formats

✔ Store (on one DF-420 D-VHS tape) up to 3.5 hours of material in HS mode (28.2 Mbps), 7 hours in STD mode (14.1 Mbps), 21 hours in LS3 mode (4.7 Mbps) or 35 hours in LS5 mode (2.8 Mbps)

The higher the *bit rate* (the preceding Mbps numbers) the less compressed the video is, and the better it looks.

✔ Record/play at 1080i, 720p, 480p, and 480i resolutions

✔ Record/play 5.1-channel Dolby Digital and DTS HDTV broadcasts for surround sound

✔ Play HiFi stereo VHS and S-VHS for Dolby Pro Logic surround sound

You'll find some other VCR features from regular VHS/S-VHS VCRs that you might recognize. For example, you might find VCR Plus+, which allows an average human being to record a certain show at a certain time more easily. With this system, all you have to do is punch in a special code that is listed next to the show you want to record in your paper's TV listings (or in *TV Guide*). More advanced versions of VCR Plus+ (called Silver and Gold) allow you to localize your VCR Plus+ settings for your area.

VCR Plus+ is handy, but once you've used a slick DVR interface (such as TiVo's), you'll never want to go back to using even the finest VCR system! Of course, you can only record HDTV on a TiVo if you buy a $1,000 DirecTV TiVo system.

At the time of this writing, we could only find one or two D-VHS products left on the market (and we looked a lot!). The most popular and prominent is JVC's HM-DH4000U (see Figure 13-2). As an indication of the market acceptance of D-VHS thus far, the list price for JVC's *only* D-VHS model (and JVC invented/promoted the D-VHS!) is $999.95, but you can get it at Crutchfield (which has competitive but not hugely discounted prices) for $599.99 — 40 percent off! So if you want one, we'd advise you to move quickly, because these might not be around much longer.

D-VHS DIGITAL HDTV RECORDER

Figure 13-2: JVC's HM-DH4000U has the D-VHS market just about cornered!

Part V
Monitor Madness

The 5th Wave
By Rich Tennant

"There's a slight pause here where the PS2 sequences your DNA to determine your preferences."

In this part . . .

Who said HDTVs had to be all about watching DVDs and high-definition satellite signals? We sure didn't. In fact, there are all sorts of other ways to get high-definition signals on and off your HD system — so many that we created this special part to talk about them.

We'll discuss all the high-definition implications of your PS2, Xbox, Nintendo, or other gaming console, including noting which consoles allow you to play DVDs on your HDTV. (Now, *that's* double duty.)

For example, we'll talk about what to do with your camcorder if you want to view your home movies in 1280-x-720-pixel mode (and what to think about when you crave a new high-definition camcorder for those really fine-looking shots of your kids in pumpkin-head Halloween outfits).

Then the fun really begins, as we discuss the wide range of gadgets that are popping up to accentuate your HDTV purchase. We'll talk about souped-up PCs designed to support your HDTV-powered home theater, video jukeboxes that make watching another movie truly point-and-click, and how to take a bath with your HDTV (shocking isn't it . . . oh, bad pun!).

We wrap with a discussion of how to take advantage of your HDTV from all over the home, with tips, tricks and guidance on home networking. Pat and Danny wrote the book on this, well, actually two books — *Smart Homes For Dummies* and *Wireless Home Networking For Dummies* — so you get a great introduction to how to access HDTV in a whole-home way.

Chapter 14

Gaming Consoles

· ·

· ·

*V*ideo game consoles and HDTVs can be great together. The *graphics engines* (the chips within the console that create video images) in today's video-gaming machines are incredibly powerful and capable of very high resolutions. While there aren't a lot of games on the market that offer higher than standard-definition (480 lines) resolution, you *can* find many widescreen, progressive-scan games that look awesome on your HDTV screen.

Understanding Consoles

If you're of a "certain age" (like us), you remember the original Atari consoles and games like Pong — crude black-and-white graphics only. If you've not kept up with video games since then, you'd be shocked at the amazingly realistic graphics that today's video-game console systems produce.

Today's market-leading video-game consoles — the Sony PlayStation 2, the Nintendo GameCube, and the Microsoft Xbox — are actually powerful computers that have been specially config-ured to play video games. In fact, these consoles are so "computer-like" that they can be set up (with official or unofficial software) for such cool non-game functions as surfing the Web, running Linux programs, and checking e-mail.

Some high-end TV aficionados look down their noses at game con-soles. Don't let this attitude (which we think is dying off) bug you at all. We think game consoles are perfectly fine for HDTV.

These are the key features of the latest game consoles:

- **Powerful graphics engines** provide true HDTV (such as 1080i format) — provided the *game* uses high-resolution video.

- **S-Video and component-video connections** bring higher-quality video *into* your television.

- **Optical drives (usually DVD)** store game software, which means more storage space for better graphics.

- **Surround-sound formats** (such as Dolby Digital) provide a truly immersive gaming experience.

If an image stays still for too long on-screen, it can cause *burn-in* — the phosphor coating on the screen stays illuminated with a certain image long enough for it to become permanent. (Hey, that's why screen-savers were invented.) Susceptible TVs — such as *plasma screens, CRT projection systems* and — to a lesser degree — *CRT direct-view* TVs — can suffer burn-in from excessive gaming. When you have burn-in, you always see a "ghost" image on-screen of whatever has become burned in. Certain images tend to be in the same place all the time when you play a game (like your race car's speedometer, or your ammo indicator), so you could experience burn-in if you put in lots of gaming time. You can pretty much avoid this problem by simply

- Setting up your TV properly (brightness *down)*
- Not gaming for hours and hours every day

LCD (liquid-crystal display) screens, LCD projectors, and DLP TVs are immune to burn-in. They'll take as much gaming as you (and your sore thumbs!) can throw at them.

Meet the Consoles

You may have an older gaming console (or three) lying around the house — such as an original PlayStation or a Super NES system. By all means, if you'd like, connect these to your HDTV. You may just have one old game that you can't help but pull out and play every once in a while (for Pat it's the old Nintendo version of *Dr. Mario* that he just can't give up). Just keep in mind that these older consoles (typically using an RF modulator or — at best — a composite video connection) won't rock your world on an HDTV. Aside from the primitive graphics, the resolution is limited to standard-definition specs.

We focus our attentions in this chapter on the latest versions of the major gaming consoles (the Sony, Nintendo, and Microsoft consoles), because these give you the best results with your HDTV. All of these consoles have been on the market for a few years, they've become quite affordable (well under $200 each), and there are hundreds of games for each system.

The PlayStation 2

The most popular gaming console — and the one with the most games available — is Sony's PlayStation 2. As with all the consoles we're discussing, the PS2 (that's what all the cool kids call it) is really a powerful computer, based on a 300 MHz CPU (central processing unit). 300 MHz may not sound fast (compared to a 3 *GHz* Pentium 4 PC), but the PS2 has some other stuff that makes it great for games, including these goodies:

- ✔ **A powerful graphics chip:** Called the Graphics Synthesizer, this chip — not the CPU — does all the heavy lifting to create the gaming images that show up on your HDTV screen.

- ✔ **32 MB of RAM:** Computer memory to hold the actual game code being executed by the CPU and the Graphics Synthesizer.

- ✔ **A DVD-ROM drive:** PS2 games are burned onto DVD discs — you can also use this drive to play regular DVDs and CDs, and even older PlayStation (the original version) games.

- ✔ **Ports and connections:** The PS2 is loaded with ports and connections for attaching peripherals — including connectors for game controllers and USB peripherals, a slot for memory cards, and an i.LINK (FireWire) port for connecting multiple PS2s together for head-to-head gaming.

- ✔ **An optional network kit:** For the ultimate in PS2ing, you can buy a network kit that provides an Ethernet interface for connecting the PS2 to a home network and the Internet. Online gaming is the ultimate goal here — the ability to play with people thousands of miles away! (We talk more about home networking in Chapter 17.)

The PS2's graphics capabilities could theoretically support HDTV resolutions, but the system is designed for lower resolutions. With the optional component video cables — and a game that supports it — you can have progressive-scan gaming (480p) on your HDTV.

The PS2 can support Dolby Digital 5.1-channel surround sound (as long as you use a digital-audio cable to connect the PS2 to your A/V receiver or surround-sound decoder). Most games, however, provide only two-channel stereo sound.

The back of a game's packaging has the Dolby Digital logo if a game supports Dolby Digital.

Gaming with the Cube

Another popular choice among gamers (particularly the younger crowd — if you have kids, you probably already know this) is Nintendo's GameCube. The GameCube is slightly less powerful than the PS2 or the Microsoft Xbox, but it makes up for this by being the cheapest console at $99.99 retail. The GameCube is also amazingly compact (6 inches or less in all directions).

Nintendo (in what we assume is a cost-saving measure) has, as of the summer of 2004, removed component-video connection capability from new GameCubes. Many games support 480p progressive scan, but you won't get that through a new GameCube. If progressive scan is important to you, consider looking for a *used* GameCube on eBay or elsewhere.

So what's inside the GameCube? How about all this good stuff crammed into that small package . . .?

- ✔ **A powerful graphics processor:** Jointly designed by ATI (big in the PC graphics chip market) and Nintendo, this chip (codename: *Flipper*) can crank out up to 12 million polygons per second — which is a lot of video on-screen.

- ✔ **40 MB of system memory:** For storing gaming code.

- ✔ **Four ports for gaming controllers:** So three friends can come over and play head to head.

- ✔ **Two memory card slots:** For storing game "state" information, so you can start back up where you left off.

- ✔ **A proprietary optical disc that's smaller than traditional DVDs or CDs:** In other words, no double duty as a DVD player for the GameCube.

As we mentioned earlier in this section, many GameCube games can provide non-HDTV, progressive-scan signals (480p) to your HDTV — but the hardware (at least in the latest versions) is lacking. If you've got an older GameCube, you can take advantage of this progressive-scan signal by using component-video cables.

Is it a console or a PC?

We've mentioned the fact that gaming consoles are really just specialized PCs dressed up in fancy, living-room-friendly clothing. Well, a company called Apex Digital (you may not have heard of them, but they are actually one of the world's largest manufacturers of DVD players, among other electronic gizmos) has just announced a new console that's even more PC than the rest.

The Apex Digital ApeXtreme Media & Game Console, due on the market by the time you read this, is a special-purpose console designed for playing PC games (not PS2 or Xbox or GameCube games) on your TV instead of on your PC monitor.

Basically, the ApeXtreme console is a combination of a DVR (see Chapter 12), a DVD player, and a gaming console, all in one slick package. We got to see an early prototype at the Consumer Electronics show and were impressed — we still haven't seen the final version, but we're waiting with bated breath.

The big advantage of this new console is that it plays any PC game (there are tons and tons) — so you aren't locked into a certain console type when you go to choose games. All this for only $499 (list price).

You can find more information about the Media & Game Console on Apex Digital's Web site: www.apexdigitalinc.com.

If you want surround sound with GameCube games, set your A/V receiver to *Dolby Pro Logic II.* The GameCube doesn't have a digital audio connection to hook up your surround-sound decoder, so there's no support for Dolby Digital surround sound.

X (box) marks the spot

The king of the gaming consoles for HDTV is the Microsoft Xbox. That's because the Xbox is the only console that

✔ Supports true HDTV outputs (720p or 1080i):

- Just connect the picture via component-video cables, choose the right games, and you're in HDTV gaming heaven.

- The Xbox supports Dolby Digital 5.1 channel surround sound, and many games support this directly.

✔ Has (at least a handful of) games that support HDTV:

It isn't just the console that gives you an HDTV gaming experience — the game *software* must support HDTV, too. This support means loads of painstaking work for all the engineers and artists creating the game, so there are just a few HDTV games on the market as we write. We expect that this number will grow — particularly as the next generation of these game consoles reaches the market.

The Xbox, more than any other console, is built around a PC-style architecture. (This probably isn't surprising, given that Microsoft is involved.) The heart of the system is a fast CPU from the Pentium III family. The graphics are handled by a scorchingly fast chip dubbed the X Chip, built by NVIDIA (another big PC graphics-chip maker). Other key components of the Xbox include the following:

- **A built-in 8GB hard drive:** Instead of using relatively small memory cards, the Xbox has a generous-size hard drive for saving game-state information. You can also "rip" your own CDs onto this hard drive to create your own game soundtracks.

- **64MB of high-speed RAM:** Not a lot of RAM by PC standards, but considerably more than PS2 or GameCube for holding game code.

- **A built-in Ethernet port:** While networking kits are optional for the PS2 and GameCube, every Xbox includes this Ethernet port for connecting to the Internet — and to online gaming via a home network with a broadband cable or DSL connection.

- **A DVD optical drive for games, CDs, and video DVDs:** So you can play DVDs and CDs with your Xbox (keep in mind, however, that the remote control that enables DVD playback is optional).

Connecting to Your HDTV

When you connect a gaming console to your HDTV, it's decision time. Out of the box, each of these consoles usually has only

- A composite video connection (the yellow connector)
- A pair of analog audio cables (the red and white connectors)

If you're buying one of the many, many different bundle packages, you might get some different cables in your box, but it isn't likely.

Video connections

Most video games come with a yellow *composite* video connector.

If you want to get the most from your gaming console and your HDTV, consider buying a higher-quality S-Video or component-video connection, in either of these ways:

> ✔ Directly from the manufacturer of your console

> ✔ From a third-party cable vendor like Monster Cable (which sells a wide range of video console cables)

Going with composite

A composite-video connector provides the standard, out-of-the-box" connection for any of these gaming consoles. A composite-video connection is also the bottom of the video-connection hierarchy; it gives you a lower-quality picture than either an S-Video or component-video connection.

If you aren't playing any games that are specifically designed for HDTVs (that is, no HDTV games on your Xbox, no 480p-resolution games for any console), then this connection usually is adequate (though not optimal). But we recommend you consider a better connection.

An S-Video connection can provide a significant picture boost over composite video for your console, but if you ever plan on playing progressive-scan or HDTV games, you should definitely skip over the composite-video connection and move on up to a component-video connection.

If you use the composite video connection (maybe you're just a casual gamer, or the console is just for the kids and they move it around a lot and can't handle the complicated cabling), you can connect several ways: All these methods work equally well.

> ✔ You can run a cable directly into one of the composite-video inputs on the back of the HDTV set.

> ✔ You can use the front-panel input (if your TV has one).

> ✔ You can connect through your A/V receiver (if you're using one) and use its video-switching facilities.

If the game is for the kids, and they may either move it to the bedroom TV or take it to a friend's house, use the *front panel* connections on your TV — just to keep the kids from messing around with *any* connections on the back of your system.

Stepping up to S-Video

S-Video is a big step up over composite video because it separates the *luminance* and the *chrominance* (the brightness and the color information in the video signal) onto their own conductors instead of trying to cram all that information into one signal (on one conductor) and then separate it when it's inside the TV.

If you aren't playing 480p or high-def games on your console, you can probably just use S-Video to connect your console to the TV. There's a definite convenience factor at work here — your HDTV is bound to have more S-Video connections available than it does component-video connections. In fact, most HDTV front-panel connections include an S-Video connector, so you can easily plug and unplug the console whenever you want to move it around.

S-Video connectors are the biggest pain-in-the-you-know-what in the audio/video world. They're prone to *bent pins*. If you're going to let the kids make this connection, make sure you teach them how to line things up correctly before making the connection.

Depending on your equipment, you can connect S-Video the same three ways as composite video: direct to the back of the TV, to the front, or through an A/V receiver. Just pick the method that's most convenient for you.

Going component

When you really want the best for your HDTV gaming experience, use component-video cables. They give you progressive-scan video for games that support it, instead of just relying on your HDTV's internal scaler system to convert the game video to progressive scan. Component-video cables also give you higher-resolution video (at least on the Xbox) for true HDTV gaming.

As we mention earlier in this chapter, the very newest (as we write) versions of the Nintendo GameCube console have had their component-video support *removed* in a production change. You can still use component-video cables with GameCubes built before the summer of 2004, but not with the newest versions. Yes, we think this was a stupid move, too!

There's nothing complicated about using component-video cables with your console — just buy the appropriate cable system and plug it in. Keep in mind a couple of things:

- Most HDTVs don't have component-video connections on the front panel, so you must connect to the rear of the TV.

- Most HDTVs have *few* component-video connections (often only one set), so you may have to decide whether to connect your game, cable box, or DVD player to your HDTV.

If you don't have enough component-video inputs on your TV, you can route your game console through an A/V receiver's component-video switching system.

If you run component video through your A/V receiver, and you want to get progressive scan (480p), make sure that your receiver's video switching has at least 10 MHz of bandwidth; to get HDTV, you need at least 30 MHz of bandwidth. You can find out how much bandwidth your video switch has by reading your A/V receiver's manual.

When you get your component-video connection set up, don't throw away that composite-video connector that came with your console. We've heard that some combinations of HDTV and consoles *play* through component-video cables (and work well!) but require a composite-video cable for some *setup* work in the console's built-in menus and controls.

Dealing with audio

Your game system probably has two connectors (red and white) for analog stereo sound. If your system has digital surround sound, you may be able to connect it to your home-theater audio system for wall-to-wall sound.

Analog audio

With analog audio cables, you can get some nice stereo sound from your HDTV or A/V receiver. You can even — if your A/V receiver supports Dolby Pro Logic II — get some *synthesized* (that is to say, fake) surround sound.

Analog audio cables don't have the true surround sound that's encoded in an increasing number of games. So you'll miss out on some fun audio effects (like the sounds of the bad guys creeping up behind you — literally!).

Digital audio

To get true surround sound from a PS2 or Xbox console, you must upgrade to a digital audio cable. This connection is made using a *Toslink* optical cable — often sold as part of a package with the component-video cables.

There's no option for digital audio with the GameCube.

Usually, you connect this Toslink cable to one of the "digital audio in" connectors on the back of an A/V receiver. Very few HDTVs accept this connection directly to their own built-in audio systems.

For real surround sound, we highly recommend using a separate A/V receiver and speaker system. This approach gives you use of your five front, center, and rear speakers, plus a subwoofer — so you can really *feel* those explosions rock the room.

What systems do we have? Pat loves his Xbox, and Danny's kids stick with their PS2 . . . all connected with component-video and Toslink cables into our HDTVs. Nice, very nice.

Chapter 15

Camcorders

*M*ost of our friends make the big camcorder purchase decision when they have their first child. There's some sort of
parental angst about not having home movies of your kid's first
views of the world.

Well don't look now, but for people with an HDTV, the ante was just
upped, as the first HDTV camcorders are starting to hit the market.
That's right, now you can start capturing live and in vivid HDTV
color anything you want. (While your guests might be bored with
the home movies, at least they'll appreciate that they've got won-
derful high-definition clarity!).

Going HDTV for a camcorder will cost you — at the time of this
writing, there was only one consumer HDTV camcorder on the
market, the JVC (www.jvc.com) GR HD1, and that model runs
about $3,500. Still, we expect the natural progression of HDTV into
camcorders to continue; soon there will be lots more choice in the
marketplace.

In this chapter, we explore how standard SD camcorders fit into
your HDTV system, and then go into what distinguishes an HDTV
camcorder from this standard off-the-shelf fare, using the JVC GR
HD1 as a case study.

Your Run-of-the-Mill SD Camcorder

Most of the concepts covered elsewhere in this book — in terms of SD resolution, viewing SD images on your HDTV, connecting SD devices to your HDTV, and the like — also apply to SD camcorders. Whether you are using 8mm, VHS, DV, or some other analog or digital format, you can view it on your HDTV.

Connecting your camcorder to your HDTV

If you think back to Chapter 3, you have an idea about a concept we strongly believe in: a *hierarchy* among the many connection options available on most video sources (including camcorders). That is to say, there's a definite good-better-best rank order among these connections.

✔ **Going digital:** If your HDTV has a FireWire (1394) input, you may be able to use the FireWire connection on your camcorder (found on most miniDV camcorders) to connect digitally. If this works with your camcorder (and it doesn't always work), it produces the best-quality picture on your HDTV. We expect to see digital connections (either FireWire, DVI or HDMI) to become available on more and more camcorders.

Even if both your HDTV and your camcorder have FireWire connections, you may not be able to use them to connect the two together. Check your camcorder's manual — many times you can only use FireWire for connecting to a PC.

You *can* use the FireWire connection to send your video to your PC, edit it (using iMovie on a Mac, or Windows Movie Maker on Windows XP), and then burn your own DVD or copy to your D-VHS recorder (see Chapter 13). This gives you a very high-quality video source for HDTV based on your home movies.

✔ **Go with S-Video:** Most modern camcorders (particularly miniDV models) include an S-Video output. The S-Video connection provides a better picture by separating the color and brightness portions of your video onto separate cables — allowing your HDTV to display them without having to use the *comb filter* to separate these picture elements.

✔ **The composite solution:** The least attractive solution for connecting your camcorder to your HDTV is to use the yellow composite-video connection. We always recommend using S-Video over composite video, but some older camcorders simply don't give you that option.

Why the wait for HDTV camcorders?

If you've been waiting for HDTV to come to your camcorder, it's only been recently that this has been made possible, for largely two reasons:

✔ It's only been recently that the digital signal processors (DSPs) have become small and smart enough to be able to handle the immense volume of data created by the high-definition CCD imaging sensors in your camcorder. Each frame offers a megapixel of resolution or more, and this has to be processed and recorded in real time to digital video tape.

✔ It's only been recently that consumers have cared! As the 4:3 television sets get replaced by 16:9 sets, people can see HDTV in its native mode, and that means a bigger market for the camcorder manufacturers to sell to.

Some day in the future, we expect that non-HDTV camcorders will also offer the *component*-video cable connection method — this will make sense as more camcorders become capable of dealing with progressive-scan video. Right now, only HDTV camcorders can use this connection method. In terms of our hierarchy, we place component video below the digital connection methods, and above S-Video.

So while you may not know it, merely buying this book is helping get more HDTV camcorders to market! Thanks for helping out.

Enter HD Camcorders

So if you can link your SD camcorder to your regular TV, what's the big deal with HD versions? Lots of pixels, for one thing — three times as many pixels as the best offered by NTSC versions, encoded as a standard MPEG-2 stream.

Ever wonder why the SD camcorders look so poor on your HDTV? It's simply a reflection of the lower resolution of the SD stream when scaled up and viewed on your HD screen — the two were simply not meant for each other.

But lots of pixels mean lots of megabytes, too. An uncompressed 1280 x 720 BMP file can be almost 3 megabytes in size — that's just a single frame of video (1/30th or 1/60th of a second's worth of video!). That's the price of high-definition.

Of course, HD camcorders use a ton of computer horsepower to *compress* these video frames down so they use less storage space — but HDTV video still uses a ton!

Camcorders that record in HD formats are called *HDV* camcorders, after the standard they use for recording (similar to the way many standard definition camcorders are called MiniDV or Super8).

But with an HD camcorder, you'll have lots of recording and playback modes that you can choose from.

- ✔ **480i:** Just like any 480 interlaced digital-video camcorder on the market, you can create regular 4:3 30-frame-per-second video streams, with the same old standard PCM 32 kHz 4-channel (or 48 kHz stereo) audio.

- ✔ **480p:** A step up is 480 progressive-scan video at 60 frames per second (fps). Here the camera records at double the frame rate.

- ✔ **720p:** This widescreen 16:9 HD format is what you buy an HD camcorder for. Current camcorders are providing 30 fps for 720p video, but we expect in the future we'll see full 60 fps 720p.

- ✔ **1080i:** HDTV camcorders may also be capable of converting their 720p video to 1080i. This could be useful if you're connecting your camcorder to an HDTV that expects (and requires) a 1080i input — some CRT-based HDTVs work this way.

If you're confused by all the FPS and progressive/interlaced information, check out Chapter 21, where we explain it all in detail!

When dealing with the higher resolution and more detailed image of a high-definition camcorder movie and its display, everything is more noticeable. Errors such as shaky handheld shots, too much panning, and zooming too fast can make people dizzy (if not nauseated) when viewed on large display. Think about using a tripod more or bracing yourself more securely when making videos.

Other than the resolution, much of the camcorder will look familiar to you — expect the same high-speed interfaces (such as FireWire) and the same tape formats (such as DV).

Checking Out the First HDV Camcorder

While there have been specialty camcorders that offer HD, and for some time (see sidebar, "Live from outer space — in HDTV"), the first consumer-oriented and -priced product came from JVC: the GR-HD1 camcorder, which retails around $3,500 but can be had on the street (or Internet) for $2,700 or less. Figure 15-1 shows this HDTV beauty.

Figure 15-1: Record in high-def with the GR-HD1.

The GR-HD1 camcorder gives you a way to get your personally recorded 1280-by-720-pixel content to your HDTV system and even archived on your D-VHS VCRs (discussed in Chapter 13), which are also sold by JVC and can play back prerecorded D-Theater HDTV movies. JVC includes a FireWire interface and software with the camcorder that converts 720-lines-of-resolution, progressive-scan HD footage (720p) into progressive-scan, anamorphic widescreen DVDs using your PC — that's cool! Material recorded either in DV or HD format can be downconverted or upconverted as needed for playback (depending upon what format your HDTV requires).

If you have a Macintosh, you're going to be disappointed. Editing with the JVC is done in the Windows XP environment with bundled software. There are also a couple of manufacturers that sell a plug-in for HDV editing with Final Cut Pro.

You can record at 480i, 480p, or 720p, depending on your preferences. JVC supports 720p at 30 fps (actually 29.97fps, but who's counting?), and some people prefer filming at the 60fps of 480p in order to get a smoother flow of the action. The 480p format is lower in resolution, but looks smoother due to the higher frame rate, and retains the 16:9 widescreen format — so it's great for creating DVDs (which can't handle 720p anyway).

Note that when shooting video with the JVC camcorder in standard 480 DV mode, you can record to the DV tape in both SP and LP modes — as you can with your existing SD camcorder. But when you shift to recording in either the 1080i or 720p mode, the unit shoots only in SP mode. This limits how much you can tape in HD mode to a one-hour maximum per tape.

Stop the presses!

While JVC is the first with a consumer HDTV-capable camcorder, it's not the only big consumer electronics company to get into the act. As we were finishing *HDTV For Dummies,* Sony announced the HDR-FX1 – a new three-CCD design (meaning it has a separate chip for recording red, green, and blue images). The HDR-FX1 isn't on the market yet, but will retail for 400,000 yen on the Japanese market. The HDR-FX1 is expected to be available outside Japan by the end of 2004.

Audio is recorded using MPEG-2 compression at 384 Kbps for both the 480p and 720p modes. Unfortunately, the GR-HD1 does not support four-channel audio or audio overdub, as does standard DV.

To get the best possible quality, play back your high-definition recordings on a high-definition monitor. On the JVC GR-HD1, this is possible only through an analog component-video or FireWire output interface. The recorder hooks up to computers and other devices via FireWire.

One great thing about the JVC is that they did not complicate the move to an HDTV camcorder by introducing a new tape format, too (and the last thing we need is another gaggle of tape formats). The JVC uses standard Mini DV tapes to store HDTV content. If you look at Table 15-1, you can see that nothing is compromised, though by using DV tapes — there is more than enough space on the tape for a 19.7-megabit HD content stream. You're using three times the pixels of a standard NTSC DVD stream when you're in 720p mode with a speed of 30 fps; naturally you need three times the bandwidth.

Table 15-1	Video Data Rates for Various Media	
Date Rate	*Type*	*Notes*
1.151	VCD	Video CD
2.756	SVCD	Super Video CD
6.5	DVD	Average data rate of DVD
8.1	Broadcast 720p 24	HDTV: 1280x720 24fps
9.8	DVD	Maximum data rate of DVD
10.2	Broadcast 720p 30	HDTV: 1280x720 30fps

Date Rate	Type	Notes
14.1	D-VHS	STD mode: Up to eight hours of high-quality 480p or 480i
15.0	Broadcast 1080 24	HDTV: 1920x1080 24fps with progressive scan
16.9	Broadcast 720p 60	HDTV: 1280x720 60fps
18.6	**JVC GR-HD1 Camcorder**	**Minimum data rate**
18.8	Broadcast 1080 30/60	HDTV: 1920x1080 30fps with progressive scan or 60fps with interlaced scan
19.7	**JVC GR HD1 Camcorder**	**Average data rate**
25	DV Tape	Stream of independent frames similar to motion JPEG. Audio bit rate is recorded separately, consuming up to 1.5 megabits per second.
28.2	D-VHS	HS mode: Maximum quality, up to four hours of 1080i or 720p
50.0	Straight stream of NTSC I-frames	Used in DigiBeta and for studio editing work
375.0	D-5	Uncompressed format for 1080i content commonly used for studio masters
400.0	First-generation FireWire	Maximum speed

Source: Camcorderinfo.com

Future models of this and other HDVs will undoubtedly move in the same direction as SD camcorders — which includes adding three CDs for more color range in the recordings.

If you want to check out the most recent HDTV camcorders on the market, check out our site at www.digitaldummies.com or go to www.camcorderinfo.com for great reviews of the latest HDTV camcorders. And if you want to know more about digital camcorders and how to make movies, check out *Digital Video For Dummies* by our buddy Keith Underdahl.

Live from outer space — in HDTV

We're used to having ultra-nice pictures from space of the Earth in all its glory, but we bet you'd never guess that HDTV was involved. That's right, even the space shuttle crew is into HDTV camcorders. HDTV equipment flew as early as STS-95 (1998), and included a Sony HDW-700A high-definition television camcorder, wide-angle lens, battery packs, and video recording tapes.

NASA is using the high-resolution images to provide clearer pictures about life on the Space Station and to improve the documentation of space exploration.

The system is enhancing the capability of NASA scientists, researchers, and engineers to conduct their research, monitor experiments, and record the data visually. HDTV also allows the public to experience NASA's explorations more realistically by making the footage available over NASA TV.

So here's a tip from those who learned the hard way: Next time you're in space, watch out for space radiation — it can cause degradation in your Charge Coupled Device (CCD) image sensor (the silicon chip inside the camera — a rectangular array of light-sensitive cells). The degradation showed up on NASA pictures early on, as a loss of several pixels on images taken on board. The camera lost between 5 and 15 pixels per day. So they created a self-correcting camera that replaces the bad pixels with an average of the luminance and chroma from adjacent pixels. Problem solved. (That's your tax money at work!)

Chapter 16

Gadgets

*W*e love gadgets. If you are getting into HDTV, we expect you're no different from us (or the rest of the world).

In this chapter, we cover cool HDTV gadgets that can act as direct sources of content to enhance your video oasis. Whether it's a high-powered home-theater PC or specialty gear destined to connect to your HDTV, you can really boost your HDTV's usage with just the right accenting gear.

HDTV technology itself is only slowly starting to infiltrate the gadgets that you'd use with your HDTV system — in other words, most of the stuff in this chapter doesn't send an HDTV signal to your HDTV. There aren't many true HDTV gadgets out there yet, but we tell you what's available today and where to find it.

Home Theater PCs

While nearly all of today's PCs are *multimedia*-capable (they can display pictures, play sounds, and show video), a select few PCs can be considered *Home Theater PCs (HTPCs)* that can feed video (and surround sound audio) into an HDTV.

HTPCs are simply high-powered PCs running the Windows, Linux or Mac operating system (the majority use Windows), which have been specially configured with hardware and software that lets them operate as the DVD player, TV tuner, or even DVR (see Chapter 12) source for your HDTV.

No rigid set of rules defines what makes a regular PC into a HTPC, but here's what we recommend:

✔ **Video card:** Perhaps the most important item in an HTPC (particularly one that feeds video into an HDTV) is the video card. This specialized set of computer chips spares the computer's CPU from most of the "heavy lifting" of video processing. Both ATI's Radeon series (www.ati.com) and NVIDIA's GeForce series (www.nvidia.com) include high-end video cards that can support HDTV resolutions and support HTPC applications.

✔ **Audio card:** If you want to support the surround sound (Dolby Digital and other systems like DTS) found in HDTV broadcasts and on DVDs, you need a relatively high-end audio card in your PC. A card like the Sound Blaster Audigy or Audigy 2 (www.soundblaster.com) is a good choice.

✔ **CPU:** There isn't a hard-and-fast rule here, but you build an HTPC around a PC with a fast processor. Look for either

- A 2.8GHz or faster Pentium 4 processor

  Microsoft requires a 3GHz Pentium 4 for full 1080i HDTV playback with Windows Media.

- An AMD Athlon 64 3200+ processor

✔ **Hard drive:** You need a decent-size hard drive on any media-centric PC. Media take up room; you need enough to store such data as MP3 and other music files.

If you want an HTPC as a DVR feeding your HDTV, go for

- At least 200 gigabytes of hard drive space

- Plenty of FireWire jacks for attaching more hard drives

✔ **TV tuner:** If you want to use the HTPC as a DVR, a TV tuner card is essential. Most of the DVR hardware/software kits on the market include a TV tuner.

If you just want the PC for playing DVDs or Internet content, a TV tuner card inside your HTPC might not be important.

✔ **Software:** There's a wide range of software you might eventually want for your HTPC — such as software that turns your PC into a PVR or organizes your media library.

The most important software may be a video-utility application such as Powerstrip (http://entechtaiwan.net/util/ps.shtm) that helps you perfectly match the resolution of your video card output to your HDTV.

A great place to get advice, see the results of people's projects, and generally dig into the topic of HTPCs is on the AVS Forum Web site (www.avsforum.com). Check out the section titled "Video Processors and HTPC" for all the info you could ever want.

An easy way to get into the HTPC game is to buy a new *Media Center PC* with the Microsoft Windows XP Media Center Edition operating system. Media Center PCs meet stringent minimum hardware requirements that allow them to work as HTPCs right out of the box. The big advantage of an MCE PC (besides having the hardware checklist all "X'd" off) is the software — the Media Center software provides a really nicely integrated experience for such home-theater functions as DVD playback, DVR functionality, and TV-watching.

MCE PCs are optimized for connections to widescreen HDTV sets, but not all MCE PC functions (such as the DVR function) support HDTV resolutions today. We expect that HDTV will be an even bigger part of the experience in future MCE PCs.

If you're interested in a lot more information about MCE PCs, check out another one of our books — *Windows XP Media Center 2004 PCs For Dummies.*

Hi-definition fun

If you've never visited AtomFilms (atomfilms.shockwave.com), you are missing out on one the greatest sites on the Web. Atom Films, which was bought by Shockwave, specializes in creating and buying all sorts of animated and live-action independent film shorts and making them available on its site. Under each section, you can find the top five shorts for that genre. Be sure to check out the comedy section. The most popular comedy when we visited was *Survivin' the Island,* a funny look at what happens to *Survivor* stars when they return from the island.

But what's really distinctive about AtomFilms is its new Hi-Def service (http://atomfilms.shockwave.com/af/spotlight/collections/hidef/). Hi-Def is a free service that really brings together broadband and HDTV. It automatically delivers AtomFilms hits to your computer for near-DVD quality viewing on your HDTV. Each week, Hi-Def will deliver three cool films using the idle bandwidth on your computer system. Films arrive automatically — all you have to do is sit back and enjoy them in their full-screen, Hi-Def glory. Films are yours to watch as many times as you want until they expire two weeks later. When older films expire, new films automatically appear to take their place. This is simply too much fun.

Video Jukeboxes

It isn't unusual to find movie junkies with thousands of DVDs. Organizing and accessing these DVDs can be troublesome at best, and the more you have, the harder it is to find one movie that you really wanted to watch. Enter video jukeboxes, gallantly riding to the rescue.

Jukebox 101

A DVD video jukebox does what you'd think — stores massive numbers of DVDs so that you can watch them on your HDTV in a more organized and accessible way (which is great when you want to watch movies all over the house).

The entry-level video jukebox is the *megachanger,* also called a DVD *carousel.* These DVD devices can store 300 or more DVDs and play any of them (usually one at a time). Some DVD changers can even control a *second* changer (doubling the capacity, for example, from 400 discs to 800)!

A DVD jukebox needs the following features (which become more and more essential the as the jukeboxes get bigger):

- ✔ A *library* function that can keep track of what disc is in which slot

- ✔ An easy way to access that library function, such as

  - • An OSD (on-screen display) that you see on your HDTV

  - • An LCD screen display on the jukebox itself

- ✔ An easy way to feed information into the library

The *easiest* way to enter information is a PC keyboard, *not* the remote control. Depending on the jukebox, you connect a keyboard in one of two ways:

- • Directly to the jukebox

- • By linking to a PC (usually a USB connection)

Check whether the device automatically looks up disc-title information on the Internet — it can save you a ton of data-entry time.

A video jukebox needs the same basic DVD player features that we cover in Chapter 11, such as

✔ Progressive scan

✔ Support for other disc formats, such as CD-R and MP3 CDs

✔ Support for new *audio* formats such as SACD or DVD-A

✔ A full range of digital audio outputs and HDTV-friendly video outputs (such as component video)

Jukebox 301

For advanced students (okay, for those with more advanced budgets!), you can leave behind the DVD jukebox, and move into the hard-drive-based video jukebox.

These devices are really more properly classified as *media servers* — computer-based devices that store various media (such as video, audio, and photos) on a hard drive.

Media Mogul

A company named Molino (www.molinonetworks.com) is launching a pair of these souped-up video jukebox servers: the Molino Media Mogul and Media Mogul TB. These servers use large hard drives (a terabyte in the case of the Mogul TB — 1,000 gigabytes — you know, *large!*), and a sophisticated user interface (UI) designed by Molino to control the system.

Media Mogul servers can "import" content from these sources:

✔ Any DVD (or CD) you own. The hard drive saves a perfect copy (including any special menu items) that you can play whenever you want, without needing to find the DVD itself.

✔ Digital pictures from your digital camera and home video from your camcorder (via USB or FireWire cable connections — see Chapter 3 for more on these connections).

The Moguls can connect to your HDTV via DVI video connections (as well as standard analog video connections) — keep in mind, however, that the DVDs you copy onto the Media Mogul aren't HDTV sources! The 300-gigabyte version of the Media Mogul costs $995, and the 1-terabyte version costs $2,995.

Kaleidescape

The Kaleidescape System is a truly high-end media server solution (www.kaleidescape.com).

The big advantage of the Kaleidescape system is its modular, multi-room support — but you'll pay accordingly. A Kaleidescape system can *start* at over $27,000 for the base system.

The Kaleidescape System is similar in functionality to the Media Mogul systems, only everything is cranked up a bit. In particular, the Kaleidescape is high-definition-ready, and supports 720p and 1080i HDTV. This multiroom system consists of

- ✔ A central server with interchangeable hard-drive cartridges
- ✔ A movie player that connects to your HDTV

  As you expand your budget and desires, you can add movie players connected to HDTVs (or any TV) in other rooms.

- ✔ A DVD reader

The whole system connects via Ethernet cabling (see Chapter 17 for details on Ethernet LANs). It can use a broadband (cable or DSL modem) connection to the Internet to update content (like the movie-guide service) and upgrade the system's software.

Do the Roku

Roku Labs (www.rokulabs.com) has a fabulous product, the HD1000 ($299). It's an ultra-sleek, high-definition, digital media player that's the first to support HDTV. You can use the Roku HD1000 to view stunning digital photos, music, video, and art on your HDTV. Tired of having all your relatives crammed around your PC monitor to see your latest pictures? Show them the snaps on your HDTV. Want to show your artistic side during a party? Why not display the latest art gallery quality photos of Renaissance art? What a Renaissance man (or woman) you are!

The HD1000 is simply a worthwhile addition to your HDTV, even if just for access to the Art Packs.

The HD1000 is better than standard non-HDTV digital media players, hands down. The Roku HD1000, unlike standard video players, does two things to ensure that the image quality is not compromised:

- ✔ It shows the image at HD output resolution. Its 1080i resolution is equivalent to about 2 megapixels, which is a huge improvement over the ⅓-megapixel output of standard non-HDTV digital media players.

A standard media player, when it displays photos from your digital camera on your TV, scales down the picture by dropping out information to fit the standard-definition format of 720 x 480 pixels. On a practical basis, this reduces the quality of a picture from your 3- or 4-megapixel camera to a mere ⅓ megapixel — a huge reduction in quality. This is why standard-definition media players always make your pictures look wimpy, even if they're hooked up to a high-definition display.

✔ It stores *all* the pixels of the source image so you can zoom and pan the image without losing clarity.

The Roku HD1000 also plays digital music files over your home network, so you can finally enjoy that growing MP3 library on your home stereo system.

But what shakes Danny's cerebrum is the Gallery Collection Art Pack, which includes images of famous art and compelling photos to turn your HDTV into a showcase. Roku offers themes like The Classics, Nature, Aquarium, Space, Holiday and Clocks on Compact Flash and CD-ROM. Way cool!

To hook up the HD1000 to your HDTV, just run a video cable from the HD1000 to your TV (Component cables are included; VGA, S Video, and composite are also supported). You can connect to your home LAN via an Ethernet port. It even supports Wi-Fi to get to your Internet connection too, if you'd like to get there that way.

We find nothing to complain about with this product.

There are a lot of other cool media-adapter products on the market (like the Prismiq Media Player, found at www.prismiq.com, or the Omnifi, found at www.omnifi.com) that can bridge the gap between your PC and any TV so you can send audio, video, and photos. Not much to say about these here; none of the popular models currently support HDTV.

Looking for HDTV Gadgets

Given the pace of change in the industry, here are the best places to check out if there are any new HDTV gadgets that artificially stimulate your fancy:

✔ www.digitaldummies.com: Our companion site for all our books. We update our site with information about updates in HDTV topics, including new gadgets you should consider.

✔ Geeky gadget sites with the latest innovations:

- www.gizmodo.com: This is Pat's favorite site.

- www.engadget.com: If this looks similar to Gizmodo, that's because the site's editor used to be editorial director at Gizmodo.

✔ www.ehomeupgrade.com: This site tracks various developments in digital media gear, including streaming video and digital media servers — two areas where we expect a lot of change in HDTV.

Chapter 26 covers other accessories that you can tie into your HDTV system.

Making HDTV Wonderful

If you're looking to build an HDTV-capable HTPC, one step you can take is to invest in ATI's new HDTV WONDER video card (http://www.ati.com/products/hdtvwonder/index.html), or one of ATI's ALL-IN-WONDER systems. The HDTV WONDER is an add-on to your existing graphics card, and for just $199 it gives you all you need to turn your desktop PC into an HDTV receiver — including a QAM/VSB tuner that can pick up any of the over-the-air digital TV formats (both high-def and plain old digital TV) as well as old-fashioned analog TV broadcasts and non-scrambled cable TV channels. The system also includes software that lets the system function as a PVR (for both analog and high-def TV shows!), a remote and even an antenna for picking up HDTV broadcasts.

If your system doesn't have a powerful enough graphics card (you need at least 64 Mbytes of RAM and DirectX 9.0 support), you can also consider one of the ATI ALL-IN-WONDER packages, which combine the HDTV wonder with one of ATI's graphics cards.

Chapter 17

Home Networking

*W*e have a saying, "No HDTV is an island." Okay, so that isn't really true — we don't actually have a saying, but if we did have one, that would be it. Why? Because your HDTV is too big an investment not to get the most you can from it.

Indeed, that's a major theme throughout this and the other books we've written — if you are going to spend so much money on your entertainment system, at least make sure you can access it from elsewhere in the home to get the full return on your investment.

Your HDTV investment is not just the HDTV itself, but also your A/V receiver, DVD, DVR, CD jukebox, and all your other audio/video gear. That's a lot of money.

In this chapter, we start with an introduction to the concept of home networks, and give you an overview of the different physical kinds of networks you may have (or want to install) in your home. Then we get into detail about how home networks can complement HDTV — by feeding media into your HDTV from the rest of the home, and by providing a "backbone" that lets you access all the video "source" devices located in your home theater or HDTV viewing room, using a TV located anywhere in your home.

Layering Your House

The ideal HDTV-ready Home of Tomorrow (or maybe Next Thursday) has several of these communications networks that you can use with your HDTV:

✔ *Telephone lines* that bring your telephone service into the house and distribute it to outlets in multiple rooms

✔ *Coaxial cable* that distributes cable or satellite TV

✔ *Home computer* networks — known to techies as *local-area networks* or *LANs* — that connect computers to a shared Internet connection and devices (such as printers)

Home computer networks can use *wired* or *wireless* technologies.

✔ *Security* networks that link such devices as smoke detectors and infrared receivers to protect your home from fire and break-ins

✔ *Electrical power* distribution to outlets throughout the house may not dispense with wires, but you don't have to add any . . .

"Wait," you say, "That isn't a communications network, is it?" But yes, Watson, it can be — we explain in this chapter.

Building a home network in your home is all about these logical communications layers in your home, and being able to hop from one layer (such as your computer network) to another layer (such as your coaxial cable network). You can use these different home network *elements* (like a computer LAN) to get content to and from your HDTV viewing room and the rest of your house, too!

The Center of Your HDTV Net

If you're planning to hop from one layer to the other, it's best to have a central point where all your in-home networks come together. We call this the wiring *hub*.

Usually, the wiring hub is where such communication paths as your telephone line, cable connection, and alarm system *enter* the home.

What sort of connections might you have at your wiring hub? Well, try these on for size:

✔ **RG6 coaxial cabling** — for cable and satellite signals

✔ **Cat-5e/6 UTP cabling** — for telephone and data signals

✔ **Wireless (2.4/5.8GHz)** — wireless backbone for voice, data, security, and other applications

✔ **Audio cabling** — for your whole home audio system

✔ **Intercom cabling** — for your whole home-intercom system

✔ **Remote-control wiring** — for your whole-home remote-infrared network, so you can use your remote controls around the house. (Many newer, RF-based remotes forego such wiring requirements.)

✔ **Home-automation wiring** — for your X10 and other home-control protocols over electrical or specialty wiring

✔ **Alarm wiring** — for your fire and burglary detectors and alarms

At the wiring hub, you are best served to install a *distribution panel.* A distribution panel is a cabling nirvana where you can

✔ Connect all your home's cables

✔ Distribute signals throughout the home

Distribution panels are common in new homes. More than 80 percent of new homes have such systems. If you don't have one, consider it. In most instances, you can wire your home for next generation services for $2,000 to $5,000, and really get the most from your home's computing and entertainment investment.

If you're considering building out your home network, you need our other books on these essential topics for any digital home:

✔ *Smart Homes For Dummies* covers how to design the infrastructure in your home to move all these digital bits from point A to point HDTV.

✔ *Wireless Home Networking For Dummies* applies the latest wireless network technologies to your home network.

Getting Signals to Your HDTV

Half the HDTV home-networking equation revolves around getting video content from different parts of your home into your HDTV. This could mean bringing in content from the Internet, playing content that's stored on your PC's hard drive, or simply getting TV signals from an antenna, satellite dish or cable-TV feeder line into the room where your HDTV lives. Heck, you could even be like Danny and hook up a wireless (Wi-Fi) camera to your system and keep an eye on the kids in the backyard on your HDTV.

Tapping into PC/Internet content

PCs can be great sources of video content (and content created as other media, such as still pictures and music) for your HDTV system. This content can come from a couple of sources:

- ✔ Within your own home — such as home movies from your camcorder, content you've copied onto your PC from CDs and DVDs, and pictures from your digital camera

- ✔ Online sources via your Internet connection

In either case, you need two network elements to get media from your PC into your HDTV:

- ✔ A *media adapter.* It connects to your home network, "reads" media stored on your PC, and converts the media it finds to a format that your HDTV can read.

- ✔ A *computer network* or *LAN* (local-area network) connection between the PC and the media adapter.

Chapter 16 covers PCs, media adapters, and other PC-related gear that helps you get computer content into your HDTV.

Using a wired LAN

The traditional way to create a home LAN is the "wired" alternative — stringing together a LAN with a special type of cabling known as *UTP* (unshielded twisted-pair).

UTP cables are basically souped-up phone cables. Eight individual conductors are twisted together inside a single cable jacket.

UTP cables are classified by their capabilities — these classifications are known as "category" classifications. You should choose UTP cables classified as either Category-5e or Category-6.

A LAN using this kind of cabling is often called an *Ethernet* network (Ethernet is the *protocol*, or underlying logic, that makes the network work). Ethernet *NICs* (or network interface cards) are pretty much standard in all desktop and laptop PCs, and can also be found in media adapters and other devices that help connect PCs and HDTVs.

Wiring up an Ethernet LAN is really pretty simple — at least the *logical* layout of how wires are connected is pretty simple. The hard part is getting the wires inside your walls. At the center of your network is a device known as a *hub* or *switch* that connects

each "leg" of your network together. These legs of the network consist of individual lengths of UTP cabling, connecting a PC or other
device on the network back to the hub or switch.

Most home LAN builders use a device that combines a switch and a
router (which can receive and distribute Internet communications
to computers and devices attached to the network). These handy
widgets are known as *home routers* or *broadband routers*.

Figure 17-1 shows a simple wired LAN.

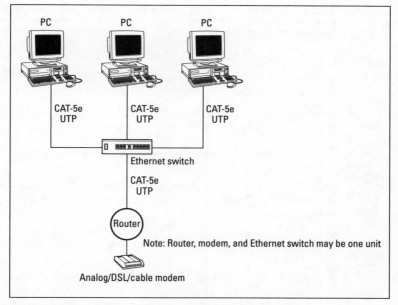

Figure 17-1: Connecting devices to a wired Ethernet LAN.

Using a Wi-Fi network

A wired LAN is a great way to connect PCs into your HDTV, but it
isn't always a convenient way to make that connection. Running
wires from your den or home office (or wherever the PC is located)
to your HDTV viewing room isn't easy (unless you're building a
new home and have open walls).

The best way to get around running these wires is to use a *wireless*
LAN system. These systems use radio transmitters and receivers
to send *packets* of Ethernet data over the airwaves instead over
wires.

When you go shopping for a wireless LAN (or *Wi-Fi*, short for wireless fidelity) system, you see more than one kind of Wi-Fi network. There are three common, standardized systems for Wi-Fi (*standardized* meaning that equipment from different vendors can work together, as long as its design is within the same standard). Wi-Fi gear is labeled with one or another variant of the "802.11" standard — there is a letter after the ".11" that designates which standard is which. Table 17-1 summarizes each of the three standards and their uses in a home LAN.

Table 17-1	Comparing Wireless LAN Systems			
Technology	*Frequency*	*Speed*	*Compatibility*	*Usage*
802.11b	2.4 GHz	11 Mbps	802.11g (at 11 Mbps)	File sharing, music
802.11a	5 GHz	54 Mbps	none	File sharing, music, video
802.11g	2.4 GHz	54 Mbps	802.11b (at 11 Mbps)	File sharing, music, video

802.11b is the most common system, but the two faster systems are rapidly supplanting it. Which system to choose depends on two factors:

✔ The big advantage of 802.11g is that it is "backward-compatible" with 802.11b.

✔ 802.11a is common, but it's less susceptible to interference because it uses a different, less crowded radio frequency.

Revving up your RF distribution

Most homes have some minimal system of *coaxial* cables in the walls carrying HDTV (and standard-definition) programming from your antenna, satellite dish, or cable-TV feed.

Running an *extra* RG-6 coaxial cable into the room where your HDTV is located (as in Figure 17-2) can pay big dividends. You can use an extra coaxial cable to create your own "in-house" TV stations that let you access all the gear that feeds into your HDTV from other parts of the house (we explain the details later in the chapter).

If you're installing a distribution panel or buying a home that has one, make sure that the panel can handle HDTV. Over-the-air DTV broadcasts use higher frequencies and have larger variations in signal strengths than do regular broadcast and cable TV stations. HDTV requires a high-quality distribution panel and amplifiers that are rated for HDTV. Panels that handle higher-frequency DTV signals are clearly marked — usually with an official DTV logo authorized by the Consumer Electronics Association (`www.ce.org`). Companies such as Channel Plus (`www.channelplus.com`) or Leviton (`www.leviton.com`) offer such amplifiers and distribution panels.

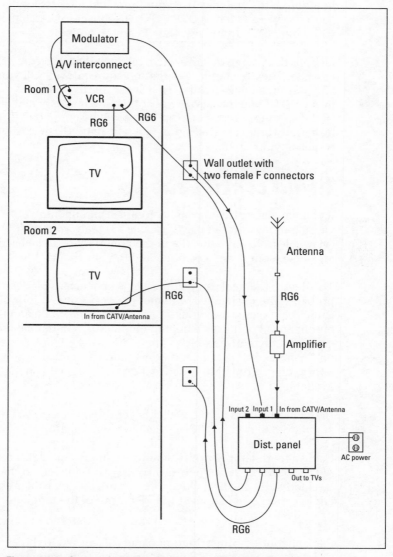

Figure 17-2: Connecting RG-6 cables to the RF distribution panel.

Sharing Signals in Your House

Getting video content *into* your HDTV from other parts of the house is just half of what home networking can do for you. You can also use a home network to share all the "local" sources that are sitting next to your HDTV.

In other words, you can use your home network to connect video source devices (such as DVRs, DVD jukeboxes, or satellite receivers) with your HDTV (and any other TV in the house).

Most of the source devices we're talking about here *aren't* HDTV-capable. They're standard-definition source devices. If you have an HDTV source device, you need a separate HDTV source device for every HDTV you own. (We expect this to change by mid-2005 when HDTV-sharing wireless set-top boxes will be common.) Today's home-networking systems don't let you share HDTV easily between rooms in your home.

Using coaxial cables

If you have an extra coaxial RG-6 cable running into your HDTV viewing room, you can use this cable to create your own TV channel (viewable from any other TV in the home) that can be used to share any video-source device connected to your HDTV.

The device that makes this possible is known as a *modulator*. You're probably already familiar with modulators — any VCR or older video game that connects to your TV via an antenna cable and is viewed on Channel 3 or 4 uses a built-in modulator. Standalone modulators can be much more flexible — while the really cheap ones only work on Channels 3 or 4, fancier models (known as *frequency-agile* modulators) can be tuned to broadcast on any channel you might have available.

HDTV modulators aren't available yet — you can only send standard-definition NTSC signals over a modulator.

Using a modulator is pretty simple — simply connect the outputs of the source device (such as a DVR or DVD player) to the inputs of the modulator (using S-video or composite-video cables and analog audio cables), and connect the RF output of the modulator to your extra RG-6 cable.

Back at your coax distribution panel (discussed in the earlier section "Revving up your RF distribution"), connect the far end of this extra RG-6 cable to the "modulator in" input. Figure 17-3 shows a modulator in action. When you have everything is connected, you simply follow the modulator's instructions to tune it to an "empty" (or unused) channel on your local channel lineup.

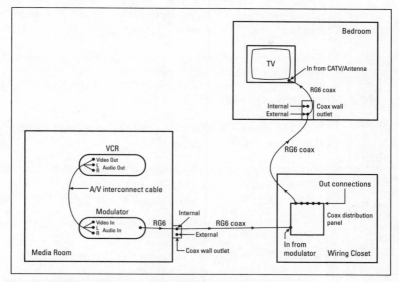

Figure 17-3: Connecting a source device to a modulator.

When your modulator is connected and tuned, you can then simply dial into the channel you've selected using any TV in the house, and watch your new "local" TV channel.

Using wireless systems

If you *don't* have extra RG-6 cabling, you can use a wireless system to distribute a video source throughout your home. Wireless video-distribution systems typically do not use Wi-Fi networking, but instead connect using their own *proprietary* wireless systems (which means you can't mix and match between vendors).

Many of the media adapters we discussed in Chapter 16 use wireless technologies (Wi-Fi) to connect PCs and Internet connections to your HDTV. These systems, however, don't send signals back from your HDTV source gear to other parts of the house.

You can find wireless video-distribution systems at any major consumer-electronics vendor (either brick-and-mortar or online). These systems aren't very expensive — for example, Terk's LeapFrog System (www.terk.com) is under $100 at online retailers.

Using UTP cabling

If you've got a Category-5e or Category-6 cable system installed in the room with your HDTV, you can use this cable to distribute high-quality audio and video (including HDTV signals, in some cases) throughout your house via a UTP-based video distribution system.

These systems aren't compatible with an Ethernet network — so if you also have an Ethernet network running into your HDTV viewing room, you need a separate run of UTP cabling to connect a video-distribution system.

UTP-based video-distribution systems range pretty widely in terms of their costs and capabilities:

- ✔ For $500 (list price) you can get the ChannelPlus SVC-10 system, which can distribute standard-definition (NTSC) S-video (and related audio signals) up to 1,000 feet over CAT-5e cabling.

  This system is *point-to-point*, meaning it can send one source device's video to only one other location.

- ✔ In the "if you have to ask, you can't afford it" category, you can find systems like Crestron's CNX-PVID8x3 "professional video-distribution system" — which uses the same kind of cabling, but can distribute up to 48 different sources, including component video HDTV signals, to up to 24 rooms!

  Crestron (www.crestron.com) is the king of custom installations of audio/video and control systems. The average installation exceeds $50,000. But if you've got the money, this can distribute your HDTV sources to any room in the home.

Part VI
Sensory Overload

In this part . . .

There's a lot to a good HDTV experience besides just a great TV set. Now, don't get us wrong, you want to start with a great TV set. And in the last three chapters of this part, we talk all about the different options you have for which great TV set to choose. We talk about front- and rear-projection systems, plasma and LCD flat panels, and CRT-powered TV sets — and the pros and cons of each approach. Despite what the salesperson in your local electronics store has told you, "everyone" does not buy the same model, much less the same type of HDTV. There are different types for different applications, and we help you figure that out in this part.

But before we get to that, we lay the groundwork for your overall HDTV adventure by talking about some of the other components (ohhh, that one's rich with punning) that drive your HDTV atmosphere.

First, we delve into the whole audio experience, both with the built-in speakers in your HDTV and with a home-theater surround-audio system. We argue vehemently that *the whole point* of going HDTV is to create a total immersion in the content, and you can't do that without real surround sound, and you can't have *real* surround sound with two Left/Right speakers built into the side of a high-definition monitor — no matter how good the speakers are. (Are we showing our biases yet?)

Then we'll talk about your HDTV room — nay, your HDTV theater — and how you can optimize that for the video and audio that comes with high-definition content. We'll advise you on placement, construction, wall coverings, lighting, and the like. Think of any Austin Powers movie and the atmosphere he creates with any swinging '60s party, and you get the idea — light and sound complement the video.

After that, we'll talk about the details of television engineering that will allow us ultimately to compare your HDTV set options. We'll make sure you are well founded in concepts like interlacing and 3:2 pulldown (and no, that's not a World Wrestling Foundation move).

Chapter 18

Understanding Audio

*O*f course, the picture — especially the *big-screen* picture — provided by HDTV content shown on HDTV-capable displays is the number-one, numero-uno attraction. But — and we think this is a *big* but — high-quality audio reproduction comes in a close second when it comes to making HDTV more of a "you're-there" experience. In other words, good audio (particularly, good *surround-sound* audio) lets you become more immersed in the HDTV experience — the audio helps make your HDTV viewing seem more like that proverbial window on the world.

In this chapter, we give you good, quick info on the confusing specifications and terms that you'll deal with when trying to understand the capabilities of an audio system. Some audio manufacturers fling bovine byproducts when they describe (dare we say *overstate?*) their audio capabilities, so you need some knowledge about what's actually what in your audio systems.

We also help you dig through the maze of surround-sound standards, to understand which speakers perform what function. We also cover the built-in audio systems found on most HDTVs.

 Chapter 19 covers *external* surround-sound systems. They're the best option for a great HDTV system. Most built-in HDTV sound systems are okay for casual viewing, but not so good when you really want to get into a movie.

Grasping Audio Basics

Before we get into any specifics about particular systems or audio components, indulge us by reading a few paragraphs about audio specifications.

Yes, we know, reading the specs themselves is bad enough — now we want you to read *about* reading the specs? Yikes! Trust us, though; we have a method underlying our madness.

Audio specs are some of the most misused numbers in the world. There isn't really any enforced standardization in how manufacturers measure and report the numbers behind their audio — we're talking mainly about audio power specs (watts).

For example, two manufacturers may both claim that their systems put out 50 watts per channel (a decent amount). Neither manufacturer is actually *lying* when it states this specification — but one system may be much more powerful than the other.

How is this possible? Well, the simple answer is that there are different ways of measuring the same thing (watts, in this case). Brand X's watts may not equal Brand Y's. Here's an explanation:

✓ **Watts:** The most basic measurement of an audio system — the number that gives you some idea of how loud the system is — is the power rating in watts. All else being equal, a system with a higher wattage rating should be able to play more loudly. Wattage is measured in watts *per channel* (or speaker).

Keep in mind that it takes a *large* increase in watts to make a truly noticeable difference in volume. To make a system play *twice* as loud, you have to increase the wattage by approximately *four* or more times.

✓ **THD:** Audio-system wattage is measured at a certain level of *distortion* (noise introduced by the audio amplifier system) called *THD* (total harmonic distortion). As audio systems are pushed closer to their limits (in terms of volume), they tend to produce greater amounts of distortion. Manufacturers can make a system seem more powerful "on paper" by measuring watts at a higher THD. Look for receivers that meet your wattage requirement when measured at low THDs, like .02 percent, rather than higher ones like 0.2 percent or even 1.0 percent.

✓ **Full-bandwidth power ratings:** Another gray area in power ratings is the frequency range at which watts are measured. The human ear can hear audio signals between 20 and 20,000 kHz. It's best if the system's power is measured across this entire range. Some manufacturers provide wattage ratings at only one

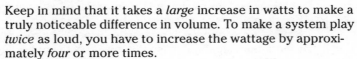

frequency (such as 1000 kHz), which can create an artificially high power rating. Try to find specifications that cover the full 20-to-20,000-kHz range to make true comparisons.

✔ **Ohms:** The number of watts a system can produce is also affected by the *impedance* (or resistance) of the speaker being driven. Manufacturers generally measure impedance at 8 Ohms, but sometimes you'll see wattage measured at lower impedances, such as 4 or 2 Ohms. It's good that a receiver can drive speakers with such low impedances (not all can), but the wattage measured at 4 Ohms is higher, so it shouldn't be compared directly to wattage measured at 8 Ohms.

✔ **Power handling:** This rating (also measured in watts) relates to the speakers in an audio system, not the amplifiers. This is simply a measure of how many watts the speakers can take before they start to shred themselves into confetti.

Power handling is not — we repeat, not — in *any* way a measure of how loud the system is. The main speaker measurement regarding loudness is the *sensitivity* of the speakers — a measure of how much volume the speakers put out with a certain wattage of input from the amplifier.

The bottom line here is to make sure that you're comparing apples to apples as you look at audio systems — either built into an HDTV, or in a separate home theater receiver system. To go back to our earlier example, two manufacturers may have 50-watt systems, but one may be measured at a limited bandwidth, at a high THD, and on a lower-Ohm impedance — and may be significantly less powerful than the other.

Surround-Sound Mania

The profusion of *surround-sound* standards is a confusing area in the audio arena. Surround sound is multichannel audio designed to produce *spatial audio cues* — sounds from all around you, in other words — relating to action on the HDTV's screen. It's usually described by the number of channels (or speakers) that a particular system uses to envelop you in sound:

✔ 5.1- audio actually has *six* channels:

- A center-channel speaker (located directly above/below your HDTV) that reproduces dialogue on your screen.

- Two front (or main-channel) speakers, which reproduce most of the musical soundtrack, plus left and right spatial cues (like someone walking into the room from one side or the other).

- Two surround speakers, located on the rear side walls of the room, that produce spatial cues *behind* you, and also provide *diffuse* (not easily locatable) sounds to help create an audio atmosphere.

- An *LFE* (or low-frequency effects) channel that uses your system's subwoofer (if you have one) to reproduce the very deep bass notes and sounds (like cannons exploding).

    The LFE channel, because it contains only a small portion of the full spectrum of audio frequencies humans can hear, is the ".1" of 5.1 (or any *x*.1 system).

✔ 6.1-channel systems add one extra speaker — a *rear surround* that is usually located on the back wall of your HDTV viewing room, and that provides an extra level of surround-sound detail.

✔ 7.1-channel systems that add *two* extra speakers, mounted on the back wall of the room.

Figure 18-1 shows a 5.1-channel surround-sound layout in a typical HDTV viewing room or home theater.

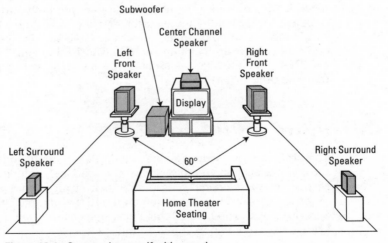

Figure 18-1: Surround yourself with sound.

Enter the matrix

A surround-sound audio signal can be created three ways:

✔ It can be encoded in a DVD or HDTV program *discretely* (each individual channel is recorded on its own channel within the audio soundtrack). This is the best way to accurately produce surround sound — when the director wants you to hear that spooky footstep *right there,* discrete surround sound gives you the best chance of hearing it there!

✔ It can be *matrixed* in along with other audio channels. Typically, matrixed surround-sound signals are mixed into normal two-channel *stereo* soundtracks. If you're listening in stereo, you don't even hear the surround-sound cues, but if you've got a surround-sound system, these "hidden" tracks are extracted from the stereo soundtrack.

✔ Sometimes there is *no* surround sound (discrete or matrixed) for an HDTV program. Perhaps the movie was filmed in 1942, when there wasn't even stereo! Well, surround-sound hardware can often create its own best guesstimate of surround sound, using a regular two-channel stereo input.

If you've got a piece of physical media (such as a DVD), you can usually figure out which surround-sound *format* (or system) is being used by looking for labels (we'll discuss the different formats in the following section). If you're watching a broadcast (HDTV or standard-definition), you might see a label or description on an onscreen program guide (or in the paper or *TV Guide*). Luckily your surround-sound *decoder* — the device that "reads" the encoded surround-sound signals and turns them into sounds you can hear — usually detects surround-sound formats automatically.

Introducing the formats

Two companies dominate the surround-sound system market: Dolby Labs (www.dolby.com) and DTS (www.dts.com). Dolby is the market leader, but both are common on DVDs and other source material. These are the most common surround formats:

✔ **Dolby Digital/AC-3:** The most common surround-sound format, Dolby Digital (also called AC-3) is part of the HDTV standard itself! Dolby Digital is a 5.1-channel, discrete surround-sound format, and in addition to HDTV programming, it can be found on most DVDs, and some digital cable and satellite TV programming.

Dolby Digital doesn't *have* to be 5.1 channels. It's possible to use Dolby Digital for two-channel stereo or even mono (one-channel) soundtracks — which is often the case for older material filmed/recorded before the advent of surround sound.

✔ **Dolby Digital EX:** As Dolby's 6.1-channel solution (with the extra "rear-surround" channel added in), Dolby Digital EX provides 5.1 channels of discrete surround sound, but then uses a matrixed system for the rear surround.

✔ **Dolby Pro Logic II/IIx:** Pro Logic II (a newer version named Pro Logic IIx is starting to appear on some audio/video gear) is Dolby's system for decoding the matrixed surround sound found on some older TV sources — like VHS VCR tapes and some stereo NTSC TV broadcasts. Pro Logic II/IIx can also create relatively realistic-sounding surround sound from true two-channel sources like CDs or stereo TV broadcasts.

✔ **DTS:** DTS is DTS's equivalent to Dolby Digital — a 5.1-channel surround-sound format. You mainly find DTS on DVDs.

✔ **DTS-ES:** DTS-ES is DTS's equivalent to Dolby Digital EX; a 6.1-channel system. Found on DVDs, DTS-ES differs from Dolby Digital EX in that at least *some* DTS-ES soundtracks use a discrete rear-surround channel. Not all do, however — look for the DTS-ES Discrete logo on the DVD case, otherwise assume you've got a matrixed DTS-ES soundtrack.

✔ **DTS NEO:6:** Not content to have equivalents to only Dolby Digital and Dolby Digital EX, DTS also has a system equivalent to Dolby Pro Logic II — DTS NEO. DTS NEO:6 takes two-channel audio input and magically creates multichannel 5.1 (or even 6.1 or 7.1) surround-sound soundtracks.

✔ **Proprietary encoders:** Some HDTVs (and other surround-sound gear) have a surround-sound system from someone other than Dolby or DTS — these systems typically provide functionality similar to Pro Logic II or NEO:6, creating multi-channel audio output from two-channel input.

So why should you care? Well, you have two issues here:

✔ Be sure your system can decode the right audio formats.

✔ Buy the versions of audio that are most compatible with your system (or else know what you *aren't* getting when you buy).

In an ideal world, your audio-playback capability would support all these formats, but you want at least Dolby Digital. Most systems support both Dolby Digital and DTS. As far as the 6.1 or 7.1-channel systems go, that's a matter of personal taste — there's very little content out there for them, so they aren't requirements.

You can tell what encoding was used to create the audio by looking on the back of the DVD or CD.

Dealing with Built-in Audio

Most — but not all! — HDTVs have a built-in audio system. TV audio systems have never been known for their audio fidelity, but you'd be amazed at how good some of these built-in audio speakers sound.

Even if you have a fancy home-theater audio system, you might find that the built-in audio system on your TV is good enough for casual viewing — such as watching the evening news, or listening in to *The Today Show* while cooking breakfast in the kitchen. In fact, there are times like this when it's probably preferable to just fire up the HDTV and use its built-in speakers, rather than warming up the mondo complicated home-theater system. When using the TV with a surround-sound system on, the TV's speakers usually assume the role of a center front speaker.

Here's what we think you should look for when you evaluate the built-in audio in an HDTV system:

- **Amp power:** While amplifier power ratings for TVs are sometimes all but useless (see the following tips), you should look for a system that puts out 10 to 20 watts per channel, if you're going to rely only on the TV's audio system.

Check out the THD (see the preceding section, "Grasping Audio Basics" if you aren't familiar with THD) and other power rating factors when you examine this number. TV wattage ratings are often measured at significantly higher THD levels than "real" audio equipment is. So that "50-watt" TV system may be equivalent to a 10-watt home-theater receiver, powerwise.

Some manufacturers claim "x" watts, but they aren't talking about individual channels, but the sum total of all amplifier totals. You certainly can't compare this number to the watts-per-channel rating of an external audio system.

- **Number of speakers:** Most HDTVs include two speakers — left and right, in other words. A few have some additional speakers that you can place throughout the room — but this feature is increasingly rare, particularly in light of the low prices of home-theater systems these days.

- **External speaker attachments:** Some HDTVs don't have extra speakers, but have amplifiers that can power external speakers. In this case, you attach your own speakers to the back of your TV with some standard speaker cables. (Another rare option.)

- **Built-in surround-sound decoder:** *True* HDTVs — those with built-in HDTV tuners — include a surround-sound decoder that can decode the Dolby Digital signal used by HDTV broadcasts.

Most HDTVs only have *two* speakers, so they won't create surround sound from an HDTV broadcast. The decoder is just there so the HDTV can turn the surround sound into two-channel audio, or feed surround channels to an external system.

✓ **Special two-channel surround modes:** Many HDTVs include special audio circuits that can help create the *illusion* of surround sound from the two speakers built into the TV. (See the sidebar titled "Creating surround sound from thin air!")

✓ **Connectivity to home-theater receivers:** If (*when*, in our opinions) you decide to move beyond the two speakers built into your HDTV, you'll want to connect your TV to a home theater receiver. If you have an HDTV with a built-in HDTV tuner, it needs a *digital audio output* to connect to your receiver. (Chapter 3 covers digital audio outputs.)

This connection is almost standard equipment for any true HDTV, but you should make sure. Look for either Toslink *(optical)* output or coaxial digital output.

Creating surround sound from thin air!

You need at least 5 (.1!) speakers for real surround sound in a viewing room, but your brain can *think* you're surrounded by speakers when you aren't.

For example, the folks at Dolby Labs have a couple of systems (which you may find in some HDTVs' audio systems) that make two speakers sound like 5.1 or more! These systems (like *Dolby Headphone,* which reproduces surround sound for headphones, or *Dolby Virtual Speaker,* which does the same with two conventional speakers) use computer horsepower to modify the sound going to your two speakers (or headphone transducers) by adding *echoes* and *delays.* These echoes and delays are designed to reach your listening position so your brain is fooled into thinking that there are more than two speakers.

Are these systems any good? They aren't bad. Dolby Virtual Speaker and SRS Labs' *TruSurround* can do a good job of fooling you. If you aren't installing a real surround-sound system, look for an HDTV with a system like these.

We don't think you can substitute these "virtual" surround-sound systems for even a small, inexpensive "real" surround-sound system (such as a $200 "Home Theater in a Box" system). But if you don't have the room or budget for 5.1 speakers, give virtual surround sound a whirl and see what *you* think.

Chapter 19

Home Theater Audio

*I*n the previous chapter (Chapter 18) we talked about surround-sound audio. Surround sound is an integral — we think, essential — part of HDTV. And the best way to get *real* surround sound is to leave behind the built-in HDTV sound system (the speakers that came installed in your HDTV, in other words) and connect a home-theater audio system to your HDTV.

In this chapter, we discuss some options for adding an external surround-sound system to your HDTV "theater." First, we talk you through the popular, all-in-one "Home Theater in a Box" solutions that include the speakers, amplification, surround-sound decoding, and often a DVD player, all in one box. These solutions are the easiest way to add surround sound to your HDTV.

We also talk about build-it-yourself systems with separate speakers and home-theater receivers. These systems take a bit more work on your part to assemble, but potentially offer better sound, since you can mix and match the "best-of-breed" pieces and parts from different manufacturers.

Boxing Up Your Home Theater

Home Theater in a Box (HTIB) systems are the *simplest, quickest,* and *easiest* way of adding surround sound to your HDTV. With HTIB, you go to the store (or shop online) and bring home a single box. Everything you need — usually including all the cables and wires that connect everything together — is right in that box, ready to go.

HTIB systems are pre-matched, preconfigured and prepared for quick and easy installation. The price you pay for this convenience is a lack of flexibility — you don't get to pick and choose individual components on their merits. HTIB systems can sound very good, but fall a bit short of slightly more expensive systems where you choose your own components (such as speakers, receivers, and DVD players).

Typically, a HTIB contains the following:

- ✔ **Amplifiers for at least five channels:** The amplifiers take the low-level audio signals recorded in an HDTV program or on a DVD soundtrack and *amplify* them electrically so that they drive your speakers. Look for at least 50 watts per channel.

- ✔ **Surround-sound decoders:** The surround-sound decoder extracts multiple audio channels from the soundtrack of an HDTV broadcast or DVD (or any audio source) and directs the audio signals to the appropriate amplifier channels.

- ✔ **An AM/FM tuner:** For tuning in radio broadcasts.

  Together, these first three items make up a home-theater receiver (discussed in the next section, "Receiving Home-Theater Sounds").

- ✔ **A DVD/CD player:** Most HTIBs include a DVD player, either in the same chassis as the amplifier and surround-sound decoders, or in a separate chassis that connects via standard audio and video cables. Check out Chapter 11 to learn more about DVD player features and specifications.

- ✔ **Five or more speakers:** HTIB systems typically include five small *satellite* speakers (they're called satellites because they hang around your subwoofer like Sputnik orbiting the earth). For 6.1- or 7.1-channel HTIB systems, expect an extra one or two speakers.

- ✔ **A subwoofer:** This speaker — usually the largest in your HTIB system — handles the lowest of the low-end frequencies in your surround-sound system.

- ✔ **Wires and cables:** Here's where HTIB makes things really easy — most systems include all the cables you need, and they're usually very clearly labeled, so it takes absolutely no brain power to hook things up.

In the section that follows, we discuss a variety of features to consider when choosing a home-theater receiver. Most of these features also apply when choosing a HTIB system.

Receiving Home-Theater Sounds

HTIB systems can be great, and they sure are convenient, but the fact is, the companies that are best at making receivers are usually *not* the best at creating speaker systems. Sure, those really huge consumer-electronics manufacturers have a lot of talented engineers on staff, but we tend to prefer speakers from specialist companies who focus solely on speaker design and manufacture. Going with your own components lets you do this mix-and-match thing.

The centerpiece of a surround-sound audio system is the *home-theater receiver* (also called an *AV receiver* or just a *surround-sound receiver*). The receiver is the centerpiece of a home-theater system. Among the receiver's many duties are the following:

✔ Decoding surround-sound signals from DVDs and other sources

✔ Amplifying audio signals and sending them to the speakers

✔ Switching between different audio and video sources

✔ Adjusting audio volume and tones

✔ Tuning in radio broadcasts, and (most importantly)

✔ Acting as the main "user interface" to the home theater

There are literally hundreds of home-theater receivers on the market these days — ranging from under $200 to well over $4,000. When we're shopping for a receiver, here are the items we make sure to check off on our list:

✔ **Power:** We like to look for receivers with at least 70 watts per channel, 90 or 100 if our HDTV is in a larger room.

✔ **Number of channels:** All home-theater receivers can support 5 channels of audio (see Chapter 18 for details on surround-sound channels); some support 6 or 7 channels for a 6.1 or 7.1 system.

✔ **Surround-sound modes:** Dolby Digital and DTS support is pretty much the baseline system; many receivers also support 6.1/7.1 systems such as DTS-ES. We also like to look for Dolby Pro Logic II support in a receiver, for providing simulated surround sound for mono and stereo content (such as older TV shows).

✔ **Video switching:** Video switching allows the receiver to act as the traffic cop for your HDTV system — taking all video source signals and routing them appropriately. Having a receiver that

can do video switching is a big plus if you have more components to connect to your HDTV than you have connections
available on the TV itself.

✔ **Component video support:** As we discussed in Chapter 3,
HDTV signals require at least component video connections —
but not all home-theater receivers support component video
switching. Look for a receiver with component video *bandwidth* of at least 30 MHz if you want to run HDTV through the
receiver.

The other cables used to carry HDTV, *DVI* and *HDMI*, are typically *not* supported by receiver video switching, so you need
to make those connections directly from source to HDTV, and
not use the receiver's video switching.

✔ **Video upconversion:** You'll probably have a variety of different
video sources feeding into your HDTV through your receiver.
Some use component video, others S-video, and still others
composite video (such as a VHS VCR). Video upconversion
takes all those sources and converts them to be carried over a
single cable or set of cables to your TV. For example, component video upconversion allows you to send all your video to
the HDTV using a single set of component video cables. Such a
set up makes it easier to connect your HDTV to your receiver
(you need less cables), and makes it easier to set up your
remote control and HDTV itself, since all video comes in over
a single set of cables.

✔ **Number of inputs:** You should take a look at the back of your
home-theater receiver (see Figure 19-1 for a typical view) and
count up the number of inputs (both audio and video — we
describe each type in Chapter 3). The key here is to make sure
that your receiver has enough of these inputs to accommodate
all the gear you might want to connect to your HDTV system.

It's sometimes useful to have a set of inputs on the *front* of your
receiver, for quick connections to devices such as camcorders.
You may also have front inputs directly on your HDTV; if you
have a flat-panel unit (see Chapter 23), you probably don't.

✔ **Assignable inputs:** One really cool feature found on an
increasing number of receivers is *assignable inputs.* Instead
of rigidly matching the receiver's digital-audio inputs to a
particular video input, assignable inputs can be used with
any video input on your system. This just makes things a
whole lot more flexible when you're trying to connect multiple devices that use digital audio (and there are many, such
as HDTV tuners, DVD players, D-VHS VCRs, and even game
consoles such as Xbox).

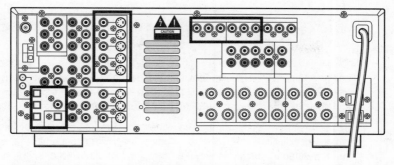

Figure 19-1: It's a jungle back here — receiver inputs/outputs.

The biggest disadvantage of skipping over the HTIB solutions (and going with a separate receiver/DVD/speaker systems) is that you have to spend a little bit more time shopping for gear (you have to pick out that receiver, a DVD player, and a set of speakers). We personally don't find this too much of a hassle (heck, we even wrote another book, *Home Theater For Dummies* on just that subject!). If you do find it a hassle, check out the HTIB systems — they pack a surprising bang for the buck.

Going with separates

Moving from a HTIB solution to your own set of surround-sound speakers and home-theater receiver is a step up, providing (potentially at least) better sound quality. The next step is to move from a home-theater receiver to *separates*.

In a separate system, the functions of the home-theater receiver are divided up among — no surprise here — separate components. A separates system usually includes the following:

✔ **Power amplifiers:** These devices provide the actual audio signal amplification that "drives" the speakers in the surround-sound system. The power amps are the *brawn* of the system.

✔ **Surround-sound controller/preamp:** The controller/preamp handles all audio and video switching — sending audio/video inputs to the HDTV. The controller also decodes surround-sound formats and "steers" audio signals to the correct channel of the amplifier (and therefore to the correct speaker). The controller is the *brains* of the system.

✔ **AM/FM tuner:** If you want *all* the functions of a home-theater receiver, you have to add a separate AM/FM radio tuner. Nope, this has nothing to do with HDTV, but we just thought we'd be complete!

(continued)

(continued)

So why are separates an improvement over all-in-one receivers? Well the difference is subtle, but there some things that separates are better at:

✔ Carrying the amplification and control/preamplification tasks in a separate chassis can be a benefit all in itself. Keeping these functions electrically separate — and using independent power supplies for each — reduces the risk of interference and noise getting into your surround sound. This is, however, a subtle difference.

✔ Separates allow you to choose the system that works best for you. Need more power (for a large room)? Get a bigger power amp. More concerned with a feature such as video upconversion? Get a fancier preamplifier/processor.

✔ Separates make it easier for you to upgrade over time (for example, when a new surround-sound format hits the market). We know lots of people who are constantly upgrading their systems on eBay, selling their old stuff, and buying someone else's *newer* old stuff.

Chapter 20

Setting the Mood

· ·

In This Chapter

▶ Creating the right atmosphere through light and sound control

▶ Optimizing walls, floors, and ceilings for HDTV

▶ Automating your HDTV theater

▶ Putting your money where you park it

· ·

When you go to the movies at your local cinema, you can't help but notice the little things they do to try to create a certain atmosphere. There are dimmed ceiling lights, moving curtains, aisle side lighting accents, and nice seats — all the creature comforts you need to watch the, er, creature feature.

However, there's a lot more going on there than you might think. The large video screens, surround-sound audio, tiered chair placement . . . these are painstakingly planned to give you an uninterrupted view of the movie content. By minimizing things that would remind you that you're in a theater, movie-theater planners create a grand illusion that you are right up there on the screen with the characters, in the thick of the action.

Buying an HDTV *can* be simply about taking the TV out of the box, plugging it in, and then watching the great picture. But why stop there — your HDTV experience could be a whole lot more exciting with the right equipment placement, lighting, sound treatments, and other nice touches that turn a living room or spare den into your true HDTV theater.

Choosing Your HDTV's Home

Turning your HDTV room into a HDTV theater is all about trying to optimize that same sense of illusion found in the movie theater. There are definitely wrong places to put your HDTV, even if you had your heart set on a particular spot next to the ficus plant. Here's a list of things to think about as you locate your HDTV:

✔ **Room layout:** You tend to get more awkward sound patterns in perfectly square rooms. The best place to put your HDTV is along the short wall of a rectangular room, preferably a wall without windows or doors on it. Fully enclosed rooms are best for sound. If you must use a room that's open to another room, consider pulling heavy curtains across an open wall when you're watching films in your HDTV theater.

✔ **Seating layout:** Many place a couch against a wall, with the HDTV in front of it. With a surround-sound-enabled HDTV theater, you want enough space behind you so the sound can get in back of you and truly surround you. So the ideal position for your HDTV seating is a location more central to the room.

✔ **Stray lighting and noise:** Stray light in a room, day or night, can substantially affect the HDTV viewing experience. The same is true for stray noise. Think about how lights from other rooms or street lighting might affect the ambience of the whole room. Listen closely to your room for regular interfering sounds, such as a clock ticking or a fish-tank pump whirring. Consider moving these devices if you can. And if the light or sound is coming externally (say, from a floodlight or a dryer or washer), consider some cheap absorptive wall coverings to shield and muffle them.

✔ **Distance to the picture:** There's a correct viewing distance and maximum viewing angle for HDTVs. (We explain it later in the chapter.)

A TV can be too *big* or too *small* for your room, and you should sit within its best *angle* of viewing.

✔ **Reflected sound and light:** Think about how video and audio signals behave in your HDTV room. Muted wall colors or irregular wall coverings — bookcases are ideal — absorb stray light. A dark gray or black room is best, or one with heavy, colored drapes. (Now you know why you see all those drapes and carpeted walls in theaters!)

Avoid mirrors, picture frames and brightly colored gloss paint. They reflect light that creates *light ghosts* on the sides of the screen.

Bare tile or wood on the floor causes acoustical reflections that mess up your *sound field* (the total sound "picture").

A good rug can absorb stray sounds that would otherwise muffle the crispness of your sound system.

It's okay for the back wall to be a little reflective — it helps build a general sound field behind your seating area.

The right distance from the picture

Obviously, where you sit to watch your HDTV is a very subjective thing, but it's a topic that some people have spent a lot of time thinking about. (What's the title on their business cards? "Deep TV-Distance Thinker"?)

HDTV screens are very different from regular TVs in one big way — because they have such high resolution, it's harder to discern the visibility of the scan lines, so you can move in closer to the screens if you want to. Crutchfield (www.crutchfield.com), always a good source for audio and video information, recommends the following distances from your HDTV:

Screen Size	Suggested Viewing Distance
30"	6.25 feet
35"	7.3 feet
40"	8.3 feet
45"	9.4 feet
50"	10.4 feet
55"	11.5 feet
60"	12.5 feet
65"	13.5 feet

Source: Crutchfield (www.crutchfield.com)

Creating the right atmosphere for your HDTV can be pretty expensive, but only if you want it to be so. In the rest of this chapter, we expose you to some simple lighting, sound, automation, and comfort concepts. Some of this is easier to do if you are building a room for your HDTV theater from scratch, versus merely sprucing up an existing room. In all instances, there's a wide spectrum of money you can spend, from the very inexpensive treatments to very involved construction. You can choose your pain.

Sounds Right?

Most people who shop for audio gear for their HDTV setup focus on how many watts this one puts out, how many interfaces that one has, and so on. What most people don't realize is that sound quality has at least as much to do with room acoustics as with the gear.

The shape of the room, floor and wall coverings, and furniture all have a massive impact on the quality of sound from your HDTV theater. The sound is getting from point A (the speakers) to point Z (your ears) through a number of intermediate points — including some sound-reflecting surfaces throughout the room.

Because you are likely to put your HDTV in an existing space, you're probably starting at a decisive disadvantage when it comes to optimizing a room for sound. We often see an HDTV in a living room that opens into the kitchen or dining room in the rear — so there is no rear wall to reflect sound. This substantially affects the aural sound field in a surround-sound system.

We're talking about sound a lot here — that's because we think one of the biggest advantages of HDTV is the built-in Dolby Digital surround sound. The picture is most important, of course, but surround sound really makes a difference in creating the "illusion of being there" that HDTV can offer.

The biggest thing to look for is vibration. Everything vibrates to a degree, including such structures as walls, ductwork, light fixtures, and woodwork. When your subwoofer belches out a low-frequency sound wave that is absorbed by the room's elements, they vibrate in reaction and establish the room's own special interpretation of the sound wave. Ultimately, the direct sound coming from the speakers combines with the secondary sounds coming from all this vibration around you; that total effect creates your acoustic summary of the action on the screen.

Half the job is simply securing everything around the room, and listening for things that add noise. Subwoofers can shake things up too much; get inexpensive isolation pads for the subwoofer's feet. If you are using an HDTV projector, consider a special mounting for it that contains the noise it adds to the equation (remember not to block the projector's fan and airflow; it puts off a lot of heat, too).

Try to avoid showing off your gear. Many enthusiasts feel an urge to put all their equipment out in the open where people can see their investment — but this merely adds more noise and heat to your room — which has to be compensated for with cooling fans and more soundproofing. Tuck away as much of this stuff as you can — out of sight, out of sound range (if possible), and away from dust. People are here to see the HDTV, not your shiny new amps!

Controlling sound in your theater

To get the most from that great digital surround sound your HDTV system is capable of putting out, take a few steps to optimize your HDTV theater room's audio "environment." Try to control two major killers of surround-sound goodness:

✔ Sounds coming into the room from other sources

✔ Unwanted sound reflections and refractions as the sound waves from your speakers interact with the room

Building from scratch (or doing a major renovation)

If you are building a new home from scratch or doing a renovation, you're lucky, because you can "optimize your environment" a lot at the structural level. You basically can design your own isolation chamber that helps preserve the impact of the HDTV content.

A few steps can optimize your new HDTV space for your theatrical presentations:

✔ **Build a new room within a room:** A room within a room can isolate and control the impact of the sound system's signals on the room itself. This room would have its own walls, flooring and ceiling to optimize the HDTV experience.

✔ **Get some distance:** When building the walls, make sure the studs of two adjacent walls (the new inside one and the existing home walls) don't touch each other. This cuts off a main path for sound to travel into and out of the room.

✔ **Add more material:** Unwanted *vibrations* (and what else is sound?) can be reduced by adding either (or both)

 • A second layer of *sheetrock* to your wall

 • More *insulation* inside the wall

✔ **Soundproof the walls:** Soundproofing is easy to install between the studs and your drywall. It should reduce both

 • Sounds traveling into and out of the room

 • Vibrations from the drywall against the studs

Acoustiblok (www.acoustiblok.com) has a helpful tarpaper-like sound-barrier material that you just tack onto the studs.

Follow the manufacturer's instructions when you apply soundproofing. Most manufacturers tell you to install soundproofing material somewhat *loosely* so vibrations can be dissipated in the material itself.

✔ **If you can, soundproof the floor:** If you have the luxury of installing an additional floor, it acts to separate the theater environment from the floor, in the same way we just discussed creating a wall within a wall. It's usually three layers:

- The bottom layer usually is sponge glued to the floor.

- The middle layer is plasterboard.

- The top layer is tongue-and-groove chipboard.

Truly advanced plans call for laying a new floor foundation with the same sort of soundproofing we discussed for the walls. If you can do that, then you have a *really* understanding spouse!

If you are putting studs onto a concrete floor, consider adding some isolators that cushion the studs from the concrete, keeping the resonance of the sound within the flooring on "your side" of the room. An example are Acoustic Innovations' IsoBloc Isolators (www.acousticinnovations.com), which raise standard studs (such as 2x4, 2x6, and 2x8) a quarter of an inch off the floor, for more tactile bass effects with the lower frequencies.

✔ **Remember the ceiling:** If you have (or plan) a suspended ceiling, check into special *spring mounts* so your ceiling doesn't rattle. Companies such as Kinetics Noise Control (www.kineticsnoise.com) have special kits for suspension ceilings to isolate sound.

If you skip this step, you'll regret it the first time you play any movie involving earthquakes or explosions.

If your HDTV home theater is in a basement with a concrete floor, consider adding a *moisture barrier* coating between the concrete and your flooring surface (such as a rug).

Upgrading your existing room for HDTV

If you aren't rebuilding your home, there are still lots of ways you can provide a much higher-quality sound experience within your existing infrastructure.

The following ideas apply whether you are building from scratch or just prettying up a room for your HDTV:

✔ **Apply wall sound control panels:** With your drywalls complete, you can mount sound-control panels on the walls to help control refracted and reflected sounds. Excessive reflected sound tends to blur the sound image and mar the intelligibility of voice tracks. Sound-control treatments cut down on the amount of sound bouncing around the room, enhancing low-level dialogue and environmental effects delivered over today's high-quality audio systems. You typically place these at your speakers' first reflection points (typically the sidewall boundaries and rear wall behind the main listening position).

Speaker *reflection points* are easy to find: Have your spouse or a friend move a mirror along each side wall while you are sitting in the spot where your couch or theater seats will go. When you can see the speaker through the mirror, this is your first reflection point, and a great spot for absorptive material.

Acoustic Innovations (www.acousticinnovations.com) has really nice solid hardwood frames in mahogany, cherry, walnut, and oak stains for its Maestro line of panels. Kinetics Noise Control (www.kineticsnoise.com) offers a Home Theater Absorption Kit that also contains special midwall decorative and absorptive panels that take care of sound reverberation.

Consider the corners of the room: Kinetics also makes special triangular corner panels engineered to absorb the low-frequency bass sound that tends to gather in corners. Acoustic Sciences (www.acousticsciences.com) has a neat product called the Acoustical Soffit, which plays the dual role: It helps control low-end bass response while it hides wiring inside a hidden raceway. If you don't have space to run wiring, you can kill two birds with one stone with products such as the Acoustical Soffit.

✔ **Remember your open spaces:** If you have open areas in your space such as a big entryway on one side of the room, or large glass windows or mirrors, consider putting heavy sound-absorbing drapery to cover these — this can help even out the sound field in the room. You can make your own draperies using heavy velvet or other drapery material from your local crafts shop, or buy some online at any of the home-theater supply sites. In a pinch, any drapes are better than nothing.

Other companies also offer lines of acoustical suppression and enhancement gear. Check offerings from companies such as Auralex (www.auralex.com) and Owens Corning (www.owenscorning.com).

You can spend as much as you want to soundproof your HDTV environs, from a couple hundred dollars to more than $50,000. You should add at least some soundproofing, especially if this is a new space without much else in the room. There are some very reasonable soundproofing treatment packages for your room; they also add a professional look to the finished room.

Lights Everyone, Action!

As with the sound in your room, controlling lighting is critical to maintaining the right atmosphere for your HDTV environment. Too much here, too little there, stray light on one side of the room — any and all of it is noticeable once you start using your HDTV enough.

You can spend as much or as little as you want on home-theater lighting. We cover $5 dimmer switches and the $300 lighting systems, but what you do is merely a reflection (no pun intended) of what you want to spend on your HDTV viewing room.

Controlling lighting in your HDTV theater

Adding lighting to your room is the type of project that you can get carried away with, we'll warn you right now. Unlike the preceding sound treatments, lighting can get fancy — for example, fiber-optic starfields on the ceiling (see sidebar, "The stars at night are big and bright . . .") — and it can be fun — like your own pre-show, er, show!

We'll start with the low-investment options upfront. For $5, you can get a dimmer switch from Lowes, Sears Hardware, or Al's Corner Drugstore. If your room has a central light, you can create the same low-light theater you encounter at the cinema; you can see where you're putting your soda and popcorn, but keep the aura of a theater.

For now a heck of a lot more, though, usually around $200 to $300, a lighting control system can give you control of lights within your HDTV room (and the whole rest of the house). These have a wall-mounted keypad, a remote control, or both; you can turn on, turn off, brighten, dim, or otherwise control each light in the room. Some packages use radio-frequency signals to communicate with the light jacks; others use X10 (see the sidebar "What's X10?").

While it's not *impossible* to find dimmer systems that work with halogen or fluorescent lights, any dimming light control will work with standard *incandescent* bulbs.

Why would you want anything more elaborate than a dimmer switch? Well, think about this scenario: When your guests arrive to watch a movie or the NBA finals in HDTV, your lighting is in Arrival mode ("house" lights are up, the bar is brightly lit, accent lights are on). When the show is ready to begin, you hit a button to enter a flashing-light period, then lighting highlights the seating and walking paths. When you're ready to start watching the movie, another button dims the lighting altogether while highlighting on the screen — then it fades, the curtains part, and the show starts.

We talk more about how to do things like this in *Smart Homes For Dummies* and *Home Theater For Dummies*, but without too much effort or construction, you can add some very effective and professional lighting effects to your home.

You can get single-room lighting systems from players such as Leviton, Lightolier Controls (www.lolcontrols.com), LiteTouch (www.home-touch.com), Lutron (www.lutron.com), Powerline Control Systems (www.pcslighting.com), Vantage (www.vantageinc.com), and X10 (the company @md www.X10.com).

"The stars at night are big and bright..."

You don't have to be deep in the heart of Texas to see stars on your HDTV room's ceiling. In fact, you can buy neat fiber optic kits to add all sorts of effects to your HDTV theater. Companies such as Acoustic Innovations (www.acousticinnovations.com) offer fiber optic kits to:

✔ **Cover your ceiling:** To make your ceiling look like a nighttime sky, you can apply fiber optic starfield ceiling panels to your existing ceiling. The panels both make your ceiling more fun and offer acoustical correction.

✔ **Cover your walls:** You can install infinite starfield panels that are made of multiple polycarbonate mirrors and fiber optic lighting and create an illusion of infinite depth. Pretty neat if you are into psychedelic midnight shows!

✔ **Cover your floor:** You can get fiber optic carpeting that's filled with tiny points of light (50 fiber optic points per square foot, to be precise).

✔ **Cover your windows:** You can get classic, velvety, fiber-optic curtains that can twinkle, too.

Heck, if you put all these in one room, you can probably think about charging admission for the room itself, not to mention the movie!

Each of the fiber-optic kits come with a dimmer, speed control for twinkle, and an on/off switch. You even get to choose between a twinkle or color-change wheel. Gosh, if these were the only decisions we had to make in life!

What's X10?

X10 is important if you're talking about installing lighting control in your home without having to run a lot of new wiring. *X10* is the dominant protocol for controlling (turning on, turning off, and dimming) electrical devices such as lights and appliances — through your home's electrical lines. You can find X10 gear from a bunch of companies, including Leviton (`www.leviton.com`) and Stanley (`www.stanleyworks.com`), and X10 Wireless Technologies (`www.X10.com`).

Using X10 could not be simpler. You basically plug in an X10 wall *module* — a small, box-shaped device no bigger than the "wall-wart" AC transformers that power telephones and answering machines — right into the wall outlet. Then the appliance (a lamp, for example) plugs into the module.

X10 *controllers* send their control signals from the controller over your power lines to every outlet in the house. When the controller finds the module it wants, it changes that module to the desired state (such as "Off"). The following figure shows a limited X10 network.

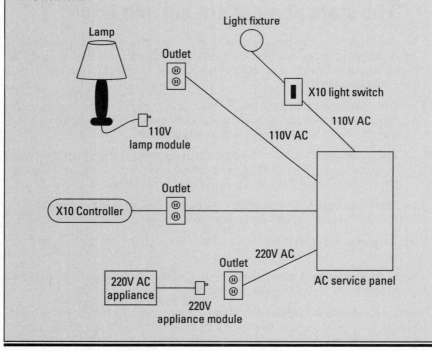

Lutron, for example, has a Home Theater Package geared for existing homes; it includes four dimmers, a tabletop master control, and an infrared receiver so users can control lights from their favorite universal or learning IR transmitter. Lutron also has a more expensive system that sports RadioRA, a wireless, whole-home lighting-control system that uses radio frequency technology instead of power lines for signal communication. With either of these packages, you can create the lighting scenarios mentioned earlier in this section. These packages range in price from $1,000 to $1,500.

The motorized shade/drape kits for your HDTV screen or windows are simple to install — if you have ever installed a drapery rod, such a kit is basically the same thing with a motor. One kit is the Makita Motorized Drape System ($600, www.smarthome.com) is a self-contained system for your HDTV theater drapes. It comes with the tracks and mounting for the system — you just add the drapes. You can also get motorized window-treatment hardware from manufacturers such as BTX (www.btxinc.com), Hunter Douglas (www.hunterdouglas.com), Lutron (www.lutron.com/sivoia), and Somfy (www.somfysystems.com).

 A great resource for ideas about lighting and room control systems — and your home theater in general — is the magazine *Electronic House* (www.electronichouse.com). It's inexpensive, available on newsstands, and has helpful hints about how to approach your projects from lots of different angles (including lots of home-theater-specific articles).

Part VII
Geek Stuff

"The picture is so sharp, you can see the Hot Wheels logo on the cars that the monster is throwing across the city."

In this part . . .

HDTV technology is, depending on your enthusiasm, either enthralling or mind-numbing. The chapters in this part lay out exactly how the pictures get on the screen.

Chapter 21

TV Engineering 101

*B*ack in Chapter 1 we gave you a 20,000-foot fly-by view of HDTV (and TV technology in general). The deeper you get into the HDTV world, the more complex and complicated the whole technical shebang gets — you go from HDTV-versus-analog into arcane discussions of lines of resolution, pixels, scanning systems, fixed-pixel displays, scaling systems, and more.

It's enough to make your head spin, trust us. So, in this chapter, we give you the deeper details that we glossed over back in Chapter 1. We explain the similarities (and differences) between lines and pixels, and we give you a firm grasp of horizontal-versus-vertical (it's not as easy as it sounds, believe it or not). We also provide more details about interlaced and progressive-scan pictures, and give you a primer on frames, fields and film-versus-video. It'll be fun, so read on!

Lines and Pixels

Okay, we admit it. When we first got into the whole video/TV/HDTV thing, we thought we had a good grip on the concept of "lines of resolution" (or "scan lines") — and pixels, and any other metric used to describe *resolution* — the level of clarity and fine picture detail on your screen.

But we were wrong — all the horizontal this and vertical that, and lines and pixels were just too much to keep up with at times. Add to this the fact that *horizontal* resolutions can sometimes refer to lines that run *vertically,* and vice versa. It's a big mess.

Going with lines

The oldest, and most common, way of describing a TV's resolution comes from the world of CRTs (cathode-ray tubes — see Chapter 24) TVs. These TVs create an image by shooting an electron beam at the screen and moving *(scanning)* from left to right and top to bottom. The lines that result are referred to as *scan lines.*

We're talking about the capabilities of the TV itself here — program images have their own resolutions. For example, an HDTV program may be 720p or 1080i — numbers that refer to the total scan lines contained in the program material.

A TV's *vertical resolution* (the metric most commonly referred to) is the number of lines that the TV can display in the vertical direction. Think of a stack of pancakes (with each pancake representing a line moving — horizontally; yes, it's confusing — across the screen). The total number of pancakes in the stack (lines on-screen) that the TV can display is its vertical resolution. See Figure 21-1 for a picture that represents this.

Don't forget! Vertical resolution measures the number of lines that run horizontally across your screen!

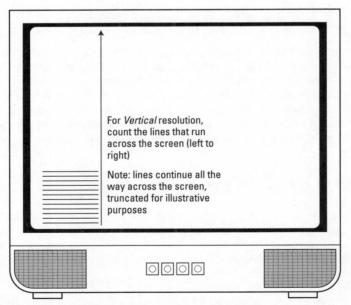

For *Vertical* resolution, count the lines that run across the screen (left to right)

Note: lines continue all the way across the screen, truncated for illustrative purposes

Figure 21-1: Vertical resolution — think of pancakes (but don't get too hungry).

We're sort of breaking the rules by describing "scan lines" and vertical lines of resolution as being the same thing. They're related, they're similar, they look/feel/smell alike, but *technically* scan lines are the lines within a TV program — and vertical resolution refers to how many lines the TV hardware can show. We think it's a lot easier to envision if you think of that electron gun scanning across the screen first, so we presented it that way.

If a TV can't display at least 720 vertical lines of resolution (720 delicious pancakes), then it isn't an HDTV-capable TV.

Now take a single one of those lines/pancakes, and look across it horizontally. Each TV can display a limited number of individual picture elements (for example, dots of alternating colors) along that line and still keep it legible. This limit is the *horizontal resolution* of the TV. Figure 21-2 demonstrates horizontal resolution.

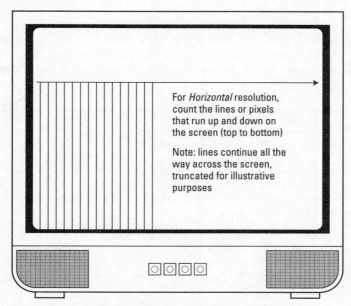

For *Horizontal* resolution, count the lines or pixels that run up and down on the screen (top to bottom)

Note: lines continue all the way across the screen, truncated for illustrative purposes

Figure 21-2: Counting horizontally to get horizontal resolution.

Picking pixels

While scanning and lines are the gist of what happens in an old-fashioned CRT TV (even a fancy HDTV version of a CRT TV), many newer TV technologies are *fixed-pixel* displays. What this means is that these TVs, by nature of their physical design, don't have an electron beam moving across a screen, but instead have thousands (or millions) of individual picture elements (pixels) that light up to create the picture. These displays are "fixed" because (unlike a CRT) they won't let you change the number of pixels in the display by re-aiming or refocusing an electron beam.

LCD and plasma flat-panel TVs (Chapter 23), and LCD, LCOS and DLP projection TVs (Chapter 22) are all fixed-pixel displays.

In a fixed-pixel display, resolution is determined by simply counting up the number of pixels in a vertical stack (from top to bottom or vice versa) — this is the vertical resolution — and across the screen (left to right or vice versa) for horizontal resolution.

The resolution of a fixed-pixel display is usually written out or referenced as "horizontal resolution x vertical resolution." For example, Pat's favorite Sony LCD rear projection TV has a resolution of 1386 x 788 pixels — 1386 pixels in the horizontal direction, and 788 vertically.

If you're used to setting up the resolution of a computer screen, you've already got this concept down. 1024 x 768 (the most common PC-monitor resolution setting) is just 1024 horizontal pixels x 768 vertical pixels.

The magic million (or so)

Instead of just comparing vertical (or horizontal) lines of resolution, it can be very instructive to compare HDTV to other TV sources (such as DVD or analog NTSC broadcasts) by comparing *total* pixels. That is to say, by doing the math, and multiplying horizontal by vertical.

A typical analog NTSC broadcast, for example, might show 330 horizontal pixels by 480 vertical pixels, for a total 158,400 pixels. An anamorphic widescreen DVD zooms up to 345,600 pixels. But even this number pales next to the potential of over *two million* pixels in a 1080i HDTV picture. Even the lower resolution 720p HDTV picture hits 921,600 pixels.

If you haven't seen it yet, you'll be simply amazed at what a million (give or take) pixels on-screen means! It's a huge difference from even the best DVD picture.

Remember that pixels aren't the only determinant of picture quality — things like screen size (and therefore how close together the pixels are), pixel shape, color accuracy, brightness, and more all work together to determine picture quality. But resolution is perhaps the most important factor, and HDTV really rules when it comes to res!

As with the lines of resolution, it's the vertical number of pixels that is most commonly referred to, and which is essential when deciding if a TV can be called an HDTV. A fixed-pixel display must have at least 720 vertical pixels to be called HDTV-capable.

The horizontal and vertical resolution in a fixed-pixel display is that display's *native resolution.* A TV signal (whether from broadcast or a recording) must be converted to this native resolution to be displayed on the fixed-pixel screen. The device that performs this task is the *scaler*, which we describe in detail in Chapter 6.

Defining standard and high resolutions

Knowing all these lines of resolutions and numbers of pixels can be an interesting intellectual exercise (if you simply *gotta* know all the stats of your HDTV), but you really need some sort of frame of reference to understand what's good and what's not.

We've already mentioned here that 720 *vertical* lines of resolution (or pixels) is the baseline requirement to reaching HDTV nirvana, but you probably want to know more details (otherwise, you'd have skipped this "techie" chapter, right?).

Table 21-1 shows the most common types of TV signals, and the TV resolutions required to show them in full detail.

If you're not familiar with the *i* and *p* suffixes in Table 21-1, we explain them in the following section, "Scanning and Interlacing."

Table 21-1	TV-Signal Resolution Requirements		
TV Signal Type	*Horizontal Resolution*	*Vertical Resolution*	*Pixels*
NTSC (analog broadcast)	330	480	158400
NTSC (DVD)	720	480	345600
SDTV (480i)	640	480	307200
EDTV (480p)	852	480	408960
HDTV (720p)	1280	720	921600
HDTV (1080i)	1920	1080	2073600

Vertical resolution gets all the attention, but horizontal resolution is important, too — it's not like your eyes can't see sideways as well as they can up and down, right? The whole HDTV industry, however, is a bit "looser" with horizontal resolution than it is with vertical — *very* few HDTVs can reproduce the full 1920 horizontal pixels/lines that 1080i HDTV signals can reproduce, so horizontal resolution is often de-emphasized in marketing materials and in articles about HDTV. You *do* want as much horizontal resolution as you can get, but you shouldn't dismiss an HDTV just because it can't reproduce 1920 horizontal lines of resolution.

Scanning and Interlacing

Resolution doesn't fully define and explain HDTV. HDTV signals (and HDTV TV hardware requirements) are also defined by their *scan type*. Traditional analog CRT monitors, designed for NTSC programming, use *interlaced* scanning; many (more modern) HDTV designs — including many CRT-based HDTVs — use a system called *progressive scanning*.

Fixed-pixel displays are inherently progressive — that's just the way they work. If an interlaced signal is fed into a fixed-pixel display, the display uses its internal scaler (Chapter 6) to convert this signal to a progressive-scan one.

Fields, frames, and your TV

Before we get into details about interlaced-versus-progressive-scan, there are two key concepts to understand:

- **Frames**: Frames are complete, full-screen images that make up an instantaneous portion of the HDTV picture. In other words, in a 720p HDTV signal, a frame is all 720 lines of the picture. NTSC images are transmitted at 30 frames per second, while HDTV images are transmitted at either 30 or 60 frames per second.

- **Fields:** Fields are simply *half* of a frame — the first field within a frame consists of the odd numbered scan lines (1, 3, 5, and so on), while the second field consists of the even numbered scan lines.

When a TV signal (whether it's NTSC or HDTV) is transmitted one field at a time, that signal is *interlaced.* Interlaced TV video has 60 fields per second (actually 59.94 fields per second, but everyone rounds up!) — these 60 fields are equivalent to 30 frames per second (fps). Figure 21-3 shows how a frame is divided into fields.

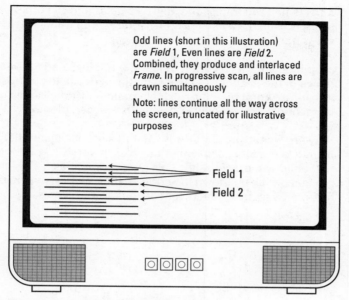

Odd lines (short in this illustration) are *Field* 1, Even lines are *Field* 2. Combined, they produce and interlaced *Frame*. In progressive scan, all lines are drawn simultaneously

Note: lines continue all the way across the screen, truncated for illustrative purposes

Field 1
Field 2

Figure 21-3: Frames and fields, oh my!

In a progressive-scan signal, both fields that make up a frame are sent simultaneously, allowing the TV to create the entire frame at once. In an interlaced system, only half of the picture is "drawn" on the screen at once, but the short time period between fields (60 fields per second!) means that your eyes see the two fields as one.

So what's better, interlace or progressive? Well, first we should answer by saying that there's nothing really wrong with interlaced video — it's not an inherently bad thing. Having said that, progressive-scan video has a smoother, more natural appearance. Interlaced video tends to have a slightly flickery appearance, at least when compared head to head with progressive-scan video.

Doing the pulldown

ATSC TV signals (standard- and high-definition) operate at either 30 or 60 fps. The film cameras that most movies (and some TV shows) are shot on operate at 24 fps.

These differing frame rates make it a difficult to convert movies to DVDs (or other formats) for standard-definition television viewing — there's not an easy way to evenly divide 24 and 30. The answer is to create an interlaced video recording (such as a DVD) that alternately repeats half the frames of video as 3 *fields* and the others as 2. This process, called *3:2 pulldown* turns the 24 film frames into 30 TV video frames.

3:2 pulldown is a clever bit of math, but it can lead to *artifacts* (flaws in the image), because sometimes two fields from different frames are combined into a single frame — these frames may not be identical, so you may end up with a jagged picture.

Why are we talking about this? Mainly because many HDTVs include a circuit called something like "reverse 3:2 pulldown" or "3:2 pulldown processing". This circuit examines a TV signal and recognizes 3:2 pulldown in action — then it does some wizardry and removes the artifacts that 3:2 pulldown creates.

Chapter 22

Projecting a Good Image

*W*hen most people think of HDTV, they think (in a Homer Simpsonesque interior-monologue voice), "Mmmmmmmmmm, big screens." And when most people think big screens, they think of *projection* TV systems. Projection TVs offer the most bang for the buck in the HDTV world (and in the TV world in general) — the biggest screen for the fewest bucks.

In the past, projection TVs (or at least most of them) also offered a lousy picture. But fear not, those days are behind us now. Projection TVs can also offer world-class HDTV pictures to go along with their large screens and relatively low price tags.

In this chapter, we talk about the two main kinds of projection systems (front-projection systems, which use a separate screen, and all-in-one rear-projection TVs). We get into the underlying technology — many different systems used to actually project the picture in a projection TV, each with pros and cons. Finally, we give you the lowdown on screen systems for front-projection TVs.

Projection HDTV Design

As we've already hinted, there are two distinct formats for projection HDTVs:

▶ Rear-projection systems encapsulate the projection system and the screen into a single chassis. These systems beam the TV picture onto the *back* of the screen.

✔ Front-projection systems consist of two pieces: the projector itself, and a separate screen. A front-projection system beams the image onto the *front* side of the screen.

Rear projection

Most projection HDTVs sold today are rear-projection TVs (RPTVs). The biggest advantage of the RPTV (compared to front-projection systems) is simplicity — all the pieces and parts are in one chassis, just as they are in an old-fashioned CRT (tube) TV (see Chapter 24 for more detail on these TVs). So with a RPTV, there's no lens adjusting and focusing of the picture on the screen. In fact, with the most modern RPTV systems (such as DLP and LCD RPTVs — discussed in the section titled "Projection TV Systems"), there's not even any need to "align" or aim the picture to avoid those awful "ghosty" images you may have seen on older RPTVs. Figure 22-1 shows a typical RPTV.

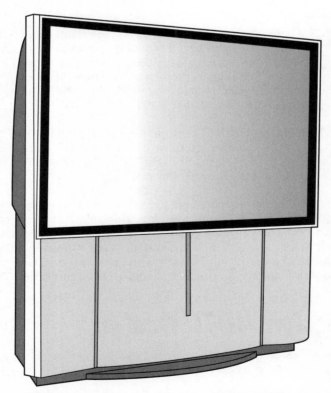

Figure 22-1: The high-def, big-screen bargain: the RPTV.

Compared to a front-projection HDTV, an RPTV

✔ Has a large (but relatively smaller) screen. RPTVs typically range from 42 to 70 inches diagonally (they can get bigger, but those are the most common sizes), while front-projection systems can fill up screens 100 inches or larger.

✔ Is generally less expensive. The most expensive RPTVs cost more than entry-level front projectors, but you can expect to pay $3,000 or less for an HDTV RPTV, while most HD-capable front projectors start at about $5,000 or more.

✔ Is easier to install and set up. Most RPTVs can be plugged in, turned on, and that's it. Even the easiest-to-install front-projection system requires adjustment in the placement of lens and chassis to focus the picture to the right size on your screen.

As far as picture quality is concerned, RPTVs (such as front projectors) can have extraordinary picture quality. The biggest factor, when it comes to picture quality, isn't so much RPTV-versus-front-projector as it is the *type* of projection system — and the quality of the individual projector.

The biggest picture shortcoming with most RPTVs revolves around the *viewing angle* of the RPTV — this describes how far from perpendicular to the screen a viewer can be and still see the picture clearly. Some RPTVs have poor viewing angles, so they're less than best when viewers are seated far to either side of the HDTV. Figure 22-2 demonstrates viewing angle.

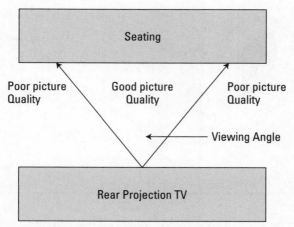

Figure 22-2: The viewing angle of an RPTV.

The other potential shortcoming of RPTVs is their size — not in terms of screen size, but rather in terms of bulkiness. RPTVs based on traditional CRT (tube) technologies can be humungous — taking more of your room than you may be willing to allow.

The newest RPTVs, which use microdisplay technologies such as DLP, are actually quite slim — barely thicker than flat-panel TVs such as plasmas.

We think that these microdisplay RPTVs are perhaps the best buy in HDTVs, combining a large screen with a slim overall package, and a picture as good as plasma at less than half the price.

Front projection

When you've really got to have a big, BIG screen for your HDTV, there's only one choice — a front-projection system. A front projector system is really a lot like being at the movies, because there's a (possibly silver) screen in front of you, and a projector behind you. In fact, some movie theaters have begun to give up film, and install DLP projectors that are really just souped-up versions of the same projectors you can buy for your home HDTV use.

Figure 22-3 shows a front-projection projector.

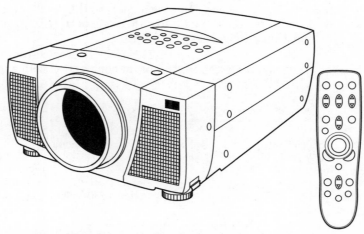

Figure 22-3: Bring the movie theater to your home.

Because you are projecting the image across the room and onto a separate screen, theoretically there's almost no limit to the size of your screen (and picture). (Check out Chapter 26, where we talk about projecting a 10-foot-by-20-foot image outdoors!)

The only real limit revolves around the brightness of image put out by the projector — because a projector must be further away from a bigger screen (and because it is lighting up a larger physical area), it needs to put out a brighter picture than it would with a smaller screen. Projector brightness is measured using a unit called *ANSI Lumens* — projectors with 500 or so are best on smaller screens (or in darker rooms), while projectors with 1,000 or more lumens are considered quite bright. Some projectors (usually those designed for PC use) are *really* bright, throwing out 2,000 or more ANSI lumens. These super bright projectors have been designed for use in brightly lit rooms, but can do double duty as projectors for very large screens.

Many front-projection TVs have no built-in TV tuner — neither HDTV/ATSC nor standard-definition/NTSC. So you need to use an external tuner, set-top box or satellite receiver to watch TV.

Projection TV Systems

When you're evaluating projection TV systems, you not only must choose between front and rear projection, you also much choose a projection technology. There are two types of projection systems:

- **Traditional projectors based on CRT (cathode ray tube) projectors.** These CRTs are specialized versions of the same tubes found in traditional TVs to generate the light projected onto the screen.

- **Microdisplay projectors.** These use microelectronic systems such as LCDs and DLP chips, in conjunction with a powerful light bulb, to project images on the screen.

Generally speaking, CRT-based systems create the best-looking picture — there's life in the old tubes yet! — but they do this at the expense of ease of use, size, and overall brightness (which relates to screen size and the kind of room a projector is used in). Microdisplay projectors trade off a little bit in ultimate picture quality (just a bit, though), but gain in ease of use (pull 'em out of the box and turn 'em on), brightness and compactness (saving floor space in your viewing room).

CRT

The granddaddy of projection TV systems is the CRT projection TV. The CRT tubes in these projection TVs are similar to those used in direct-view tube televisions (discussed in Chapter 24 — these are the regular old TVs), but with one major difference: while a direct view HDTV has a single tube, a projection CRT HDTV has three tubes, one each for red, green and blue colors.

Available in both rear and front-projection systems, the CRT is simultaneously a pain in the neck and the highest-quality projection TV you can buy. The three tubes account for the pain-in-the-neck quality — getting the picture from these three tubes properly aligned on the screen can be anywhere from slightly to maddeningly difficult to do, and accounts for many of the awful projection TV images you've seen in the past.

This process is called setting *convergence*. Some CRT RPTVs have automatic convergence systems that do a great job of setting this up for you — if you have to do it manually, it's quite difficult, and best done by a pro.

Tubes have some definite picture advantages over other systems though, as well as some additional negatives. For example:

- ✔ **Pro: best *black-level* reproduction.** Blacks are essential when you're watching a scene on TV with dark (or no) lighting and lots of shadows. CRTs are the best of all systems at reproducing black colors on screen.

- ✔ **Pro: excellent color reproduction.** Many microdisplay systems have a particular color that they can't reproduce as well as others — CRT systems don't usually face this issue.

- ✔ **Con: reduced brightness.** Compared to microdisplays (which use an intensely bright bulb to create light on the screen), CRTs put out fewer lumens.

- ✔ **Con: limited tube lifespan.** While your CRT projection TV should last years, the tubes do wear out, slowly and continuously over time.

- ✔ **Con: susceptible to *burn-in*.** CRTs create a picture when electrons hit a layer of phosphor (which lights up when struck by these electrons). Images that don't move (such as those in video games or even stock tickers) can create a permanent "ghost" image in these phosphors — effectively ruining your screen.

CRT projection TVs are the *most* susceptible HDTV for burn-in.

The finest CRT projection TVs (and the most expensive) are some of the really high-end front-projection systems. The key factor in a front-projection system is the size of the CRTs — generally speaking, the bigger the better. Most projectors have 7- or 8-inch CRTs, but a few mega-dollar units (often costing $40,000 or more) have 9-inch CRTs. These are the projectors you find in dedicated home-theater rooms of the truly rich and famous.

Rear-projection CRT-based HDTVs, on the other hand, are a relative bargain. You can easily find a high-quality, HDTV-capable RPTV with a big 50-plus-inch screen for just a hair over $1,000 — less than half of the cost of a similar-size microdisplay RPTV.

The biggest disadvantage of CRT-based RPTVs is their sheer size — these things are the behemoths of the HDTV world, particularly when it comes to the depth of the unit (how far it sticks out from the wall behind the TV). Because of the size of the tubes and the complicated mirror systems required to get the image onto the screen, some CRT RPTVs are as deep as 3 feet or more.

LCD

If you've ever used a laptop computer, cell phone, PDA, or just about any kind of electronic device with a screen, then you've used an *LCD* (or liquid-crystal display). LCDs range in size from tiny, sub-1-inch models to huge 40-plus-inch versions (check out Chapter 23 for examples of these).

LCD projection TVs tend to use LCDs from the smaller end of this continuum, often as small as 1 inch or less. Like CRT RPTVs, LCD RPTVs use not one, but three image sources — an individual LCD each for red, green and blue. Unlike CRTs, however, these LCDs don't require periodic alignment (convergence), which makes owning an LCD RPTV a much easier task for those of us who don't specialize in TV maintenance.

The small size of these LCDs accounts for what we think is the biggest advantage of the LCD RPTV when compared to a CRT model — LCD models are simply *much* thinner, closer to a flat-panel TV than a traditional RPTV in depth. So they can fit into your tight family room better than an older-tech CRT TV.

There are a couple of disadvantages of LCD projectors, however, including the following:

- ✔ **Screen doors:** LCD screens consist of a large number of sharply defined, square-shaped pixels that make up the image. When blown up to big-screen size (generally speaking, 50 inches or greater), these pixels can become visible. You'll notice this when you feel like you're looking out your home's screen door. Better LCD HDTVs avoid this syndrome, but it can show up on even the best projectors for very large images.

- ✔ **Poor black reproduction:** LCDs are a *transmissive* technology — light shines through the LCD. It's hard for LCDs to become totally black (some light leaks though), so dark scenes look more like dark gray than true black. Better LCD HDTVs have better black levels, but none match a CRT projector.

✓ **Dead pixels:** This is a *huge* deal for some folks — others won't even notice it. There are literally millions of pixels between the three LCDs found in an LCD-projection HDTV. Occasionally one of these pixels malfunctions or becomes "stuck," resulting in a visible dark or bright spot on your HDTV's screen. The real problem is that many manufacturers won't fix or replace your HDTV if you only have a few of these malfunctioning pixels — so if this sort of thing really drives you crazy, check out the warranty terms before you buy!

✓ **Lamp death:** The super-bright lamp that powers an LCD projector system has a limited lifespan — usually after a few thousand hours of use, the bulbs either dim to below usable levels or burn out altogether. While changing a bulb isn't too hard or too expensive in most cases (usually less than a couple of hundred dollars), it is a bit of a pain.

DLP

LCD isn't the only microdisplay technology vying for your projector dollars. Texas Instruments (of calculator and chip fame) has developed a system called *digital light processing* (or DLP), which is perhaps the hottest projector technology available today.

DLP systems are based on *micromirror* technology — a DLP chip (the basis of the projector) is an optical semiconductor with millions of tiny mirrors controlled by the logic portion of the chip. Basically these tiny little mirrors are individually controlled and tilted to reflect an amount of light corresponding to the picture brightness required for a single pixel of video. The angle of the mirror is changed to move from black (or close to black) — where no light is reflected onto the screen — through a whole range of grays right on up to white. Figure 22-4 shows a DLP chip.

Figure 22-4: DLP chips contain millions of tiny mirrors!

With a DLP, color is added with a separate device known as the *color wheel* — a set of red, green and blue filters arranged in a wheel that is located in the path of light reflecting off the mirrors in the DLP chip — these three colors mixed together produce the colors found in your HDTV's source material.

Some really expensive DLP projectors (like those used in movie theaters) use three DLP chips (one each for red, green and blue) instead of a color wheel. This system reproduces an even greater number of colors and tends to provide a smoother image on-screen.

DLP-projection TVs are, as we mentioned, about the hottest thing on the market today. They are very thin (some less than 6 or 7 inches — nearly in plasma territory), produce a bright, beautiful picture, with better than LCD black reproduction, and excellent color reproduction.

When you're choosing a DLP-based projection system, be sure to read the fine print. Not all DLP systems are HDTVs — some inexpensive projectors (mainly front-projection systems) use older DLP chips that don't reach HDTV resolutions. The latest DLP chip (as we write this) is the HD2+, which provides 1,280-by-720 resolution (this perfectly matches 720p resolution requirements).

Most people don't notice, but we should warn you about a situation with DLPs called the *rainbow effect*. This is caused by the spinning color wheel, and can cause a very small percentage of the population to feel dizzy, or get a headache, while watching DLP — particularly when moving their heads, or during rapidly moving scenes on-screen.

Most people *don't* have any problem with this rainbow effect — and most people who own DLP HDTVs love them and never deal with this problem at all!

LCOS

The newest projector technology on the block is called *liquid crystal on silicon* (or LCOS). Only a few LCOS HDTVs are available on the market today, but a very, *very* large company — none other than Pentium chip maker Intel — has plunged into the LCOS market and intends to put its vast scale and manufacturing expertise behind the technology in a big way.

The LCOS (which combines liquid-crystal technology — like an LCD — in a *reflective* instead of *transmissive* system) is designed to be a cheaper way of getting to really high-definition HDTVs. While most LCD and DLP systems are limited to 720p resolutions (they convert 1080i signals to 720p for playback on screen), some LCOS systems can display the full resolution of 1080i. That means that an LCOS system can display 1920 x 1080 pixels — something no other consumer HDTV system can do.

There are a handful of LCOS projection TVs available on the market today (mainly from Philips, www.philips.com), but Intel is making a huge push to get its chips into a lot of new TVs soon. We expect that LCOS projectors will be very common by the end of 2005.

Screens

While RPTVs have a built-in screen, front-projection TVs require a separate screen system for displaying the image. There's a wide range of what you can use for a screen — some people just use a white-painted wall or a tautly hung white bedsheet, for example — but most people choose a commercially built TV screen.

You can find screens that are *fixed* (permanently mounted) — these are usually the best value and the best performers. Unless you have a dedicated home-theater room for your HDTV projector, you might want a *retractable* screen (like the ones in your fifth-grade classroom, only these are often electrically powered) or *tripod* screen (that you can fold up and put in the closet when not using).

Even more important than the form are the technical characteristics of the screen. The big three are the following:

- **Gain:** *Gain* is a measure of how reflective the screen is — how much of the projector's light gets bounced back to your eyeballs. There's a standard industry reference for gain, and systems that have exactly as much gain as that reference are rated at a gain of 1. More reflective (high-gain) screens are rated greater than 1 (say 1.2), while less reflective (low-gain) screens are rated below 1 (many are rated at 0.8). If you have a CRT projector, get a high-gain screen (between 1 and 1.3, though you can go higher if needed). Microdisplay projectors can use a low-gain screen (0.8 or higher) due to their brightness.

- **Viewing angle:** Most display systems have a limited angle (from perpendicular) in which they look best. Sit outside that angle, and the picture becomes very dim. Viewing angle is inversely proportional to gain. In other words, higher-gain screens have smaller viewing angles. For this reason, you are best off choosing the lowest-gain screen that works with your projector in your room. Microdisplay projectors have light to spare, so a low-gain screen is worthwhile to gain (pardon the pun) a bigger viewing angle. Viewing angles are usually listed in a number of degrees (say, 90). Your viewing angle is half this amount (45, in this case) on either side of perpendicular.

✔ **Aspect ratio:** Screens are available in either the 16:9 widescreen or 4:3 aspect ratios. You should of course, choose the same aspect ratio as that of your projector — which, with any HDTV, is 16:9. You may want to buy "masks" that can be placed on the sides of your screen when you're watching 4:3 material (such as older, standard-def TV shows). These masks — similar to mattes for a picture frame — can help avoid any "leakage" of light beyond the edges of your picture.

Chapter 23

Thin Is In

*F*lat-panel TV technology — super-thin HDTVs that you can hang on the wall like a painting — have really taken a prime place in the popular zeitgeist. A big-screen, flat-panel plasma or LCD TV has become *the* status symbol of the '00s — not just in the living room, but in hotel lobbies, retail stores, and even in the back of cars.

There's a good reason for this mania — flat panels provide a large viewing area with almost no intrusion into your HDTV viewing room. If that was all they did, they'd be pretty cool — but they can offer a very high-quality HDTV picture as well.

In this chapter we talk about flat panels, and specifically about plasma HDTVs and LCD flat-panel HDTVs. We also discuss the pros and cons of each type, and compare them to other more traditional HDTVs.

Many flat-panel displays ship as monitors only — with neither an HDTV nor an NTSC tuner included as part of the display. Remember that you need some sort of external HDTV and NTSC tuner (or a cable set-top box or satellite receiver) to view TV on these flat-panel displays.

Loving Your LCD

We are willing to bet that you're already familiar with the LCD display. If you have a flat-panel display for your PC, a laptop PC, a PDA, a cell phone, a GameBoy — or just about any digital device with a display, you have an LCD. LCDs have been around for decades, mainly in lower-resolution formats and smaller sizes (such as phone screens), but they are getting larger all the time — and growing sharper in resolution.

We're talking about *direct view* LCDs here — in other words, LCD HDTVs where you look at the LCD display itself. In Chapter 22, we talk about projectors with LCD microdisplays — teeny tiny LCDs that are used to project a bigger image on a screen. Figure 23-1 shows an LCD HDTV.

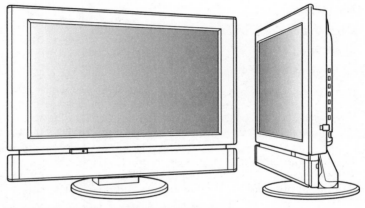

Figure 23-1: Liquid crystals put the HD in your TV.

Even as they get bigger, however, LCDs tend to be the smallest HDTVs available on the market (so they may not be big enough for a true home-theater environment). The process of "growing" LCD crystals is a manufacturing challenge, and bigger LCDs are harder to make because they have lower *yields* (or percentage of LCD crystals that can be used for HDTVs). Because of this, most LCD HDTVs on the market are smaller than 30 or 32 inches — smaller, in other words, than projectors or plasmas (discussed in the second half of this chapter), and smaller than many tube-based CRT HDTVs.

There are a few large-screen LCD HDTVs on the market these days — such as Samsung's 40-inch model — and we've seen announcements (but not yet actual products) of 50-inch and larger LCD HDTVs. Typically, however, LCD gives way to plasma once you hit the 40-inch mark, for thin TV displays.

Advances in LCD manufacturing, as well as new technologies, such as *O-LED* (Organic LED) and *SED* (Surface-conduction Electron-emitter Displays), promise LCD-like displays in ever-bigger sizes in the future. In the long run, LCD displays will be huge!

Not only will LCD HDTVs grow to huge dimensions, they'll also be huge sellers. A few reasons why include the following:

We're not including the "super-skinny-hang-it-on-the-wall" attribute here, because we think that's obvious. We can't resist mentioning it again anyway, just because it's so cool.

✔ **Excellent color:** LCDs can display millions of colors, and do so accurately (meaning the color coming off the screen is faithful to the color in your broadcast or recording). Not all flat-panel TVs (or HDTVs in general) can duplicate this color accuracy.

✔ **PC-monitor-capable:** Many LCD HDTVs can also be used as big (huge!) PC monitors. This trick is especially cool if you have one of those neat Media Center PCs we discuss in Chapter 16.

✔ **No burn-in:** HDTVs that rely upon *phosphors,* such as CRTs and plasmas, can, under certain circumstances, experience *burn-in,* where ghost images are permanently burned into the screen. LCDs are immune from this phenomenon — so feel free to play video games, watch the CNBC stock ticker, and so on with no fear.

✔ **Inherently progressive:** Unlike tube (CRT) TVs, LCDs don't rely on a scanning "gun" and *interlaced* scanning (see Chapter 21). Instead, LCDs use millions of tiny transistors that can be individually controlled by the "brains" inside the display. This means that LCDs can easily handle progressive-scan sources, such as progressive-scan DVD and HDTV.

Besides the size limitation we discuss earlier in this section, you might want to consider a few other issues before you choose an LCD HDTV:

✔ **Expensive for their size:** LCD HDTVs are great, but they're not cheap. Of course, all flat-panel TVs are relatively pricey, but LCD HDTVs typically cost more, per inch, than do plasmas. In the summer of 2004, you can expect to pay about $1,800 for an HDTV-ready, 22-inch LCD display. By comparison, an HDTV-ready, 32-inch direct-view set can be found for about half of that amount!

✔ **Poor reproduction of blacks:** Black images are among the hardest for most TVs to reproduce (CRT TVs are best at this). LCD TVs tend to produce grays, not blacks — LCD HDTVs lose out, in this regard, to both CRT and plasma TVs.

✔ **Limited viewing angle:** LCDs typically have a poor viewing angle — the angle you can sit away from perpendicular and still see a clear image on-screen. Manufacturers have been working diligently to improve this characteristic (with some success). Check the specs before you buy — most LCD HDTVs will have viewing angles listed in their specifications.

There are not only horizontal (left to right), but also vertical (top to bottom) viewing angles. If you're hanging an LCD HDTV on the wall, the vertical viewing angle may be more important to you than the horizontal — most people only pay attention to the horizontal viewing angle.

✔ **Slow pixel response time:** Another area that LCD HDTV makers are working overtime to improve is the *pixel response time* of their TVs. Basically, the individual pixels within an LCD HDTV take a slight amount of time to change color and intensity. For really fast-moving video content (particularly in a 720p picture, where every pixel can change as many as 60 times per second), an LCD TV can end up with some *artifacts* (visible flaws) where the picture from a previous frame is still slightly visible on-screen as the new one is being drawn. Typically, this isn't a huge and noticeable deal, but it's not beyond the realm of possibility that you might notice it.

✔ **Limited brightness:** The LCD is a *transmissive* system — light is shined through the liquid crystals — some of that light gets absorbed or reflected back away from the viewer. This means that LCD displays are not as bright as CRT, plasma and even some projection TVs (DLP, for example) — this could be a factor in a brightly lit room.

Everyone's Crazy About Plasma

The really hot spot in the HDTV technology market is the plasma TV. Plasma TVs combine a thin, compact chassis with a truly large (even huge) screen size, and then add beautiful high-definition pictures to the mix. For many potential HDTV buyers, plasmas really fit the bill.

A plasma screen contains literally millions of gas-filled cells (each one acting as a single image pixel) trapped between two pieces of glass. An electrical grid zaps these cells and causes the gases to ionize (and ionized gas *is* plasma — hence the name). The ionized gases, in turn, cause a layer of phosphor on the viewer's side layer of glass to light up (just as the electron gun in a CRT causes the phosphor to light up on the front of the tube).

Despite their compact dimensions (in the "depth" direction at least — many plasmas are only about 4 inches deep), plasma HDTVs are available in 42-, 50-, and even 60-plus-inch sizes. Imagine a 4- or 5-inch-deep HDTV that spans 5 feet diagonally, and you can see the instant appeal of plasma.

Other benefits of plasma displays include these:

✔ **Excellent brightness:** Plasma HDTVs are second only to CRT direct-view TVs (discussed in Chapter 24) in terms of picture brightness — plasmas don't rely on a light bulb shining through or reflecting off of something (as an LCD or DLP system does). In some ways, plasma brightness is even better than CRT's because the picture is uncannily evenly bright across the entire screen. In a CRT, on the other hand, there always is some slight (or not so slight) difference in brightness as the electron beam reaches different parts of the screen. ·

✔ **High resolution:** HDTV plasma TVs can often reach higher horizontal resolutions that CRT-based direct-view sets just can't match. The finest plasma TVs have such high resolutions (and such smooth images) that they look like nothing more than beautiful film images.

Just because it's plasma and it costs $3,000 or more doesn't make it an HDTV. We don't know of any sub-$3,500 plasma that *isn't* an EDTV *rather than an HDTV*. This will change eventually, but for now, "cheaper" plasmas will be EDTVs.

✔ **Progressive by nature:** Like LCD displays, plasma systems don't use a scanning electron beam to create a picture. Instead, all the pixels on the screen are lit up simultaneously. Progressive HDTV sources (such as 720p) and non-HDTV sources (such as progressive-scan DVD players) are displayed to full advantage on a plasma HDTV.

✔ **A wide viewing angle:** Unlike LCDs (which often have problems in this regard), plasma displays have a good picture even when you're sitting "off axis" (not perpendicular to the screen surface). This is a huge benefit for smaller rooms, where viewers may sit relatively far off to the sides of the screen, at wider angles.

Plasma's not perfect, of course:

✔ **Susceptible to burn-in:** Any system that uses a phosphor screen to display video can fall victim to the phosphor burn-in mentioned earlier in this chapter. If the Xbox is a primary HDTV source in your home, consider something besides a plasma — maybe an LCD HDTV or a rear-projection microdisplay HDTV.

You can minimize burn-in on any display by calibrating the set properly and reducing the brightness from its (usually too-high) factory setting.

✔ **Shorter lifespan:** Another phenomenon of any phosphor-based display system is that eventually the phosphors "wear out" or lose their brightness. This is a subtle and slow process, but it inevitably happens. If you've saved up to buy an HDTV to last you a lifetime, well, don't get a plasma unless your personal actuary tells you that you're close to the end of your rope.

Before you buy, check the manufacturer's specifications on *hours to half brightness* (the point at which the display is only half as bright as it was when new). For example, if this specification is 20,000 hours, and you watch the set for 6 hours a day, it will be effectively worn out in about 9 years. If you have kids, keep in mind that 6 hours a day is (comparatively) *not* a lot of time for the TV to be on every day.

- **Less-than-perfect color reproduction:** Although plasma displays are capable of producing a breathtaking array of colors, all the sets built to date have had an unfortunate tendency to make red colors look more orange than true red. If you're a huge fan of slasher horror flicks, this might take away some of your fun!

- **Poor reproduction of black:** Although plasma displays are a near equal to CRT sets in terms of absolute brightness, they fall short in the realm of reproducing black images. Most plasmas do slightly better job than LCD TVs at black reproduction, they fall short of CRTs and some projection systems (such as DLP).

Keep in mind that every type of HDTV has its pros and cons. We're certainly not trying to talk you out of going the plasma route (if your budget can handle it), just making sure that you're informed. We doubt that you'll find anything to dislike about the picture and performance of the best HDTV plasma sets — with the possible exception of the price tag!

Taking a plasma uphill

Plasma TVs rely upon a thin layer of gas (which gets ionized and turns — eventually — into the picture you see). If you think back to your chemistry classes in high school and college, you may start remembering . . . well, let's not go there. You may also, however, recall concepts like the *ideal gas law* (PV=nRT), which basically says that (all else being equal) the volume of a gas is inversely related to its pressure. Which is a nice way of saying that if the atmospheric pressure goes down (as it does at high elevations), a gas expands.

As some folks living in high-altitude cites have discovered, this expanding gas can cause the glass screen in a plasma to "bow" outward, and potentially vibrate and make an annoying noise. If you live some place high (say a mile or higher), check with the manufacturer's specs before buying a plasma set. Many manufacturers have begun to make specific design changes in reaction to this situation, and have begun selling plasmas that won't get altitude sickness.

Chapter 24

Good Ol' Tubes

*W*e've all seen (and watched) *CRT* (cathode-ray-tube) TVs — these are the traditional "tube" TVs that are in hundreds of millions of homes around the world. CRT is an "old-school" technology that's been used for TVs for more than 60 years, but that doesn't mean that there isn't still some life left in that old box.

In fact, CRTs in many ways offer the best pictures available in the HDTV world. They aren't the biggest, they aren't always the highest in absolute resolution, and they're certainly not the slimmest or thinnest HDTVs, but just about nothing can beat a CRT HDTV when it comes to brightness, reproduction of colors, and the ability to show darkly lit scenes (black reproduction).

In this chapter, we start with an overview of how CRT TVs work and what they are. We follow with the pros and cons of the CRT as an HDTV. We finish with aspect ratios. (CRTs are the one kind of HDTV screen where you often find 4:3 aspect ratios. We tell you why that is, and what you should consider if you're buying a CRT HDTV.)

All About CRTs

The CRT consists of three major elements (sure, there are a bunch of other small and significant parts — the ones listed here are the key parts):

✔ **The vacuum tube:** This is why they call a TV "the tube." The major structure inside a CRT HDTV is a large glass vacuum tube. You may also hear these referred to as *direct view* TVs — direct view as opposed to projection, meaning you watch the image directly. This tube is what makes a CRT HDTV simultaneously heavy and fragile.

✔ **Phosphor coating:** On the inside of this vacuum tube, on the opposite side of the screen from where you view the TV, is a coating of *phosphors*. These molecules emit light when they are hit with electrons — this light is the picture you see.

✔ **An electron gun:** This device at the back (or *neck*) of the tube, is the source of the electrons that pummel the phosphors on the front (screen) side of the tube. The electron gun controls this beam of electrons by aiming them (electromagnetically) into a series of *scan lines* that run across the screen.

A color CRT actually has three electron guns — one each for color of red, green, and blue. This combination of colors (often called *RGB*) can be combined to produce any color on the screen.

CRT Pros and Cons

Because CRT TVs are a mature technology, lots of folks don't give them a second (or first!) thought when they're shopping for a new HDTV. We think it makes sense to at least consider a CRT HDTV. Some reasons include the following:

✔ **Best at black:** When it comes to producing video that accurately reproduces dark scenes (such as the shadows in a horror movie when the bad guy is sneaking up on the nubile teenage girls with this chainsaw/axe/etc.). Most other HDTVs can only produce grays that approximate black — CRTs can give you something like inky black!

✔ **Work well in a bright room:** CRT tubes really blast out the light. Some of them are so bright they could almost replace the fresnel lens in a lighthouse (well, not really). What this means is that CRT HDTVs are perhaps the most suitable of all types for use in a brightly lit room. So if you're mainly watching sports in a family room with floor-to-ceiling windows, consider CRT.

✔ **Easy on the wallet:** Because the factories and machines and technologies behind CRT are old (though constantly refined), CRTs are rather inexpensive to build. HDTV CRT TVs are, of course, more expensive than their analog brethren, but still releatively cheap. You can find a 32-inch CRT HDTV for well under $1,000 — under $700 for smaller, bedroom-sized models.

You're simply not going to find a plasma, LCD or projector HDTV for that little cash (unless you're talking about a really, really tiny LCD TV).

Nothing's perfect, CRTs have some downsides, too. These are some that get us in a lather:

✔ **Size does matter:** That is to say, if you want a really big screen TV, CRT may not be for you. It's a lot easier, as an engineering and manufacturing challenge, to create a big screen TV using a projector system, or even a plasma or LCD system, than it is using a CRT. Huge vacuum tubes aren't easy to design, and they're even harder to build. If you're going to be sitting more than 10 feet from your HDTV, you should consider something besides a CRT.

Another problem with big CRT TVs is making sure that the electron beam can be accurately aimed the right places on the screen — the bigger the screen the harder it is to keep the picture consistent across the entire screen. This is just a matter of geometry — the edges of the screen are further away, and at a more obtuse angle from the gun(s) than is the center of the screen.

✔ **Portly TVs:** We're sure you've seen (and probably lifted) enough CRT TVs (be they HDTVs or just analog sets) to know this one by heart. CRT TVs, especially once their size moves beyond the 30+ inch mark, are really heavy. We mean REALLY heavy — back-breakingly, "Hey ma where's my hernia belt" heavy. They're also fat — the geometries required by the tube require a lot of depth. You'll never hang a CRT on the wall (at least not without some huge bridge girder-like stand), and it'll never be slim and sexy like a plasma or microdisplay projector system. So to give you an example, the Zenith C32V37 Direct View Integrated HDTV Television is 27 in x 36.4 in x 22.4 in (HxWxD) and 158.07 lbs ... that's typical.

The reason big CRT HDTVs are so heavy has to do with the vacuum tube. To get the tube so big without having it break under it's own weight, the glass must be extra thick, and extra thick in this case means extra heavy.

✔ **Lower resolution:** While CRT HDTVs can have extremely high resolutions, most don't reach the levels of resolution found in the best LCD or LCOS systems, or even to the resolution that the most expensive CRT-based front-projection TVs reach. Where direct view CRTs fall a bit short is in the horizontal resolution — the number of pixels running across the screen from left to right. 1080i, for example, can use up to 1920 pixels in this direction; very few CRT systems can focus in on that large a number of distinct pixels horizontally.

If you can spend some extra money, you might consider checking out a CRT-based projection TV, discussed in Chapter 22. These systems cost more, and take some extra work in terms of setup and ongoing maintenance, but they provide the picture benefits of a CRT HDTV with the big screen advantages of a projection system.

Many CRT HDTVs display images either as 480p (standard-definition) or 1080i (HDTV!). Some accept any sort of HDTV signal (such as 720p) and convert it to 1080i, but other sets require you to send a 1080i signal into the back of your HDTV. Make sure you know what your CRT HDTV requires, and make sure that your HDTV tuner, set-top box or satellite receiver can be configured to output 1080i — even when the program itself is broadcast in 720p. If your CRT HDTV has a built-in HDTV tuner, you can ignore this advice!

Navigating the CRT Jungle

If you decide to go with a CRT HDTV, you have to decide among a couple of features and attributes.

Dealing with aspect ratios

Most other types of HDTVs are available in only one aspect ratio — and that is the widescreen 16:9 aspect ratio (we discuss aspect ratio in detail in Chapter 21). Direct view CRTs differ in this regard because HDTV-capable TVs are widely available in both 4:3 and 16:9 aspect ratios.

Which to choose

In the past (and we're talking about the recent past — just a year or two ago) we often recommended that CRT HDTV buyers spend their money on 4:3 HDTVs with anamorphic squeeze functionality built in (see the next section, "Doing the squeeze"). But that was then, this is now, and we're changing with the changing times.

See, one of the big reasons we recommended 4:3 CRT HDTVs was because there simply weren't a lot of choices otherwise. But CRT manufacturers have tooled up their factory floors to produce more 16:9 tubes, and now a whole bunch of TV brands are offering 30-inch and 34-inch widescreen CRT HDTVs.

The other thing that has helped change our minds is the fact that there's simply more widescreen content available these days (and more arrives almost daily, it seems).

What we're getting at here is that we now generally recommend at 16:9 CRT HDTV over a 4:3 model for most people. We still like 4:3 models for a couple of reasons:

- ✔ They're cheaper — always a good reason if your budget (like most people's) is a bit tight.

- ✔ For the same size, you get more screen area on a 4:3 TV. This is just a matter of math: height times width = area. A 32-inch 4:3 set (which costs less) has 1 square inch more picture area than a *34-inch* 16:9 model, and probably costs 20 percent less to boot!

We recommend you get a 4:3 HDTV only if it has the anamorphic squeeze function we're about to discuss. Fortunately, just about every model we know of today *does* have this function.

Doing the squeeze

If you do choose a 4:3 television, you really should make sure that you choose one with a feature called *anamorphic squeeze* (also called *raster squeeze* or 16:9 mode). This function allows you to watch *full-resolution* widescreen content (HDTV or DVD-based) on a 4:3 HDTV screen.

Here's what the squeeze function does: While you're watching widescreen HDTV or DVD content, your HDTV re-aims its electron gun, so that the gun beams scan lines onto the screen phosphors in a 16:9 window on the screen. This re-aimed picture looks, at first glance, exactly like the traditional "letterboxed" picture that old-fashioned analog TVs use to display widescreen content — with the black bars on the bottom and top of the picture. The difference is this: With a letterboxed picture, your TV actually wastes some of its scan lines to "draw" those black bars on the screen — so your actual picture resolution is decreased, because not all of the scan lines are used to draw the picture. When an HDTV uses the squeeze function, all scan lines (usually 1080 of them) are used for the picture. The black bars are simply unlit parts of the screen. The result is a clearer, sharper picture.

Some displays have an automatic system for detecting 16:9 sources and turning on this mode, and others have a button on the remote to activate it. We like the ones with the automatic mode, of course. We're lazy.

This concept wouldn't work with most other displays — which are *fixed-pixel* displays, with a set number of pixels in each direction (horizontally and vertically). The pixels (and scan lines) on a CRT, on the other hand, aren't determined by the shape of the screen, but are instead controlled by the electron gun — which itself can be controlled and can change how it aims its beam. In a fixed-pixel display, you can change between 16:9 and 4:3, but not without leaving some pixels unused (leaving, in other words, some of your resolution on the table).

If your HDTV has this mode, make sure you've set your DVD player to "think" it's connected to a 16:9 TV. Otherwise you don't get anything from your display's anamorphic squeeze function, because the DVD player plays widescreen movies in letterbox mode, not anamorphic. (If all this anamorphic talk confuses you, check out Chapter 11, where we discuss it in more detail.)

Getting flat

CRT HDTVs will never be flat-*panel* TVs like plasmas and LCDs — there's simply too much going on behind the scenes (or in the back of the tube) to make a CRT that hangs on the wall and leaves you room for a nice game of Nerf hoops in the family room. But CRT TVs can be flat-*screen* TVs.

That is to say, the very front surface of the TV tube — the screen itself — can be manufactured so that it is truly flat.

Older CRTs (and, in fact, most CRTs still sold today) had curved screen surfaces — which made it a bit easier to design the electron gun and such. A flat screen surface, however, has the distinct advantage of not picking up nearly as many reflections or as much glare from light sources in your viewing room.

A flat screen doesn't make your picture better per se — it just makes it easier for you to enjoy that picture because you aren't squinting through some awful glare to see what's on the screen.

Part VIII
The Part of Tens

The 5th Wave   By Rich Tennant

The HDTV Cooking Channel
...and that is how we flambé.

The HDTV Travel Channel
Lapland! Land of enchantment!

The HDTV Weather Channel
Rain and flooding continue to plague the midwest.

The HDTV Comedy Channel
Welcome!

In this part . . .

Top ten lists! Don't believe David Letterman when he said he invented these — we really did. Okay, we didn't, but neither did Letterman, so there!

In this section, we suggest ten great places to go looking for your HDTV, from highly rated on-line superstores, to really informative and consultative specialty sites, to your local electronics store. There are pros and cons to buying at various places so read this before you buy anything.

When it comes to really getting the most out of your HDTV, our chapter on cool HDTV accessories is going to really hurt your pocketbook. From things that make you go bump in the night to 20-foot-by-10-foot screens for your backyard, you're going to want to get all of these items.

Finally, we discuss the most common questions we get asked about HDTV — all the things you wish you knew without asking, we're going to tell you in short form so you can look sooooo smart at your next party. "Why yes, I can tell you the difference an HDTV and 'HDTV-ready' television!" "Oh, you're sooooo smart!"

Chapter 25

Ten Places to Buy an HDTV

*B*uying an HDTV is like buying any other major piece of new electronic equipment: It can be extremely nerve-racking to try to balance the desire for the best possible features and systems with the innate fear of finding out, too late, that you've paid too much or bought something that is instantly obsolete.

Complicating this situation is the wide variety of sources and prices for HDTV systems, making simple comparisons difficult to find. HDTV systems can come packaged with other services, such as cable or satellite service. HDTVs can be sold along with a computer system or sold separately on auction sites. They are easily accessible through online buying services as well as at the traditional electronics stores. HDTV is everywhere.

There are advantages and disadvantages to each approach — there is no single answer to fit every consumer, any more than there is one HDTV to fit every living room. The good news is that with so many options, one will fit your needs, regardless of where you live.

Speaking of geography, it should be noted that our top-ten list doesn't include regional electronics mega-stores, even though these are often good places to kick the tires and do the actual purchase.

Finally, a word of warning for the first-time HDTV shopper: Be sure you know what you are buying. Many HDTVs don't come with actual HDTV receivers or tuners, meaning they are only HDTV-ready, and you'll have to buy a separate device to get HDTV programming. In some instances, a content provider such as a cable or satellite company provides the necessary tuner or receiver for free — so it helps to know where your content is coming from before you plunk down your hard-earned cash.

Crutchfield

Crutchfield (www.crutchfield.com) — an online retailer — stands head and shoulders above others for multiple reasons. In addition to the fact it is unique in offering shipping for $30 or less, free return shipping, and a 30-day trial of all equipment, Crutchfield also has a wide selection of HDTV options, organized for easy browsing. If you get to this site knowing what you want, you can get in and get out quickly.

The pitfalls of online ordering

Ordering online is, in most ways, much easier than buying from a brick-and-mortar store. It is simply much easier to find the models you are seeking and to do oranges-to-oranges price comparisons. Even if you don't wind up buying a system online, there are online price-comparison engines, such as the How Stuff Works site at http://howstuffworks.shopping.com/xGS-hdtv~linkin_id-3055268 or the excellent PriceGrabber site (www.pricegrabber.com). What's more, you can often avoid paying sales tax — which adds up very quickly on a multi-thousand-dollar item. But there are a couple of issues that any consumer must consider when attempting to order HDTV equipment online:

Shipping costs: These can wipe out savings, especially if the online vendor charges by weight instead of imposing a flat fee. Expect to pay a significant shipping fee. Most online vendors also expect you to pay return shipping, if needed, and many charge you a restocking fee if you return the unit. Most require that you return the system in its original packing and that you call and get return authorization before shipping it back. There is also a variation among vendors as to the available forms of shipping, which can drive up cost.

Installation: Installing an HDTV system may not be rocket science, but it can be a headache to non-technical buyers — and a backache to the supergeek. Remember: An online vendor simply ships the system to your house — but doesn't carry the unit inside and set it up. Some retail stores (and a few online vendors) handle setup, but that adds to system cost.

Warranty: Some manufacturers with in-home service warranties won't let you return the item after you've signed for it — it has to be repaired on premises. So check your box *very* carefully before you sign for the item.

Are they legit? This is, perhaps, the most important consideration. HDTV equipment makers want to make sure that only authorized online vendors sell their gear. The smart thing to do, before making a purchase, is to check the equipment maker's Web site to make sure your online vendor is authorized to sell that specific piece of equipment. Otherwise you could wind up without a legitimate warranty on the system — which could cost you big bucks down the road.

If you get to this site still uncertain as to what you want to buy, there is plenty of information to help you make up your mind. The site includes glossaries, background information, and an option for live online chat to get a second opinion.

Above all, Crutchfield has a great reputation for knowing everything there is to know about what it sells, and helping the customer through the whole life cycle of the purchase. This consultative approach has won legions of repeat customers. Crutchfield accepts most credit cards and checks. Orders are filled promptly, and there is online order tracking. Customer responsiveness is high, and Crutchfield also offers technical support for life — which, for newer technology such as HDTV, can be critical.

Gateway

Gateway Computers entered the home-entertainment market in some attention-getting ways. It was the first major retailer to offer a plasma-screen TV for less than $3K. It also has led the market in its launch of home-entertainment gear built around computers.

All this would be moot if Gateway didn't have a strong HDTV path, which it does. The sets connect to a PC (as mentioned), but they require separate HDTV receivers. Gateway is playing at the high end of the market versus the down-and-dirty connect-your-computer end. The company is known for solid customer support, but its return policies are similar to those of other online merchants — return authorization is required, the customer forfeits the original shipping charge and must pay return shipping, and a restocking fee is charged. The company offers extended warranty options — which might be appealing (especially to the accident-prone); the more expensive option includes on-site repairs and/or replacement for accidental damage to the TV set — and even to the remote.

 One great benefit of buying from Gateway or another computer company is that it's easier to convince the accounting department that this is a valid business expense. We're not kidding. We know one guy (shhhh!) who had a $5K use-or-lose department budget for computer gear. He basically bought a $1K computer and a $4K "computer screen." You oughta see his presentations . . .

Other computer makers are getting into the act, as the distinction between monitors and TVs are blurring — after all, a TV is just a monitor with some fancy innards for a tuner and some interfaces. As the tuner has moved out of the TV and into set-top boxes and PCs, the bulk of what is left is a monitor. Large-screen LCDs and plasma displays are increasingly common, as well as home-priced, business-class HDTV-capable projectors. So check bundled offerings from such computer dealers as Dell (www.dell.com), HP (www.HP.com), and ViewSonic (www.viewsonic.com).

What to look for when shopping online

To even a trained eye, all online shopping storefronts can look alike. How can you make sure you don't get ripped off? Well you can never be 100 percent certain, but here are some things to look for when trying to ascertain whether a site is "real" or not. You want a site that meets these criteria:

✔ Accepts major credit cards

✔ Has visible and understandable policies for privacy, returning items, for when credit cards are charged, and for getting in touch with the company

✔ Lists a physical address of some sort on the site

✔ Has a shopping-cart interface for taking and totaling orders

✔ Offers a secure server for the transaction (look for a lock in the lower-right corner of your browser to show that a secure transaction is underway)

✔ Has product-specific pages so you can make sure you are truly buying exactly what you ordered

Check out a seller's reputation. Cnet.com is a good source for this sort of information. Whenever you click on the button to Check the latest prices on Cnet, you see a listing of pricing by vendor — the "CNET certified" column tells you how well the company stacks up on Cnet's ratings, and you can click on the Store Profile listed under each store logo in order to find out specifically what is driving each company's ratings. (We never buy from any store that has received less than 85-percent positive feedback from Cnet's users. For a benchmark example, check out Crutchfield's ratings at http://reviews.cnet.com/4011-5_7-278703.html?tag=mlpmerch.)

Consider — seriously — using a service like Escrow.com to hold your money until the product arrives in one piece. Escrow.com is easy to use:

1. The buyer and seller agree to the terms and details of the transaction.

2. The buyer sends payment to Escrow.com. Payment is verified and deposited into a trust account.

3. The seller ships the merchandise to the buyer, knowing that the buyer's payment is secured.

4. The buyer accepts the merchandise after having the opportunity to inspect it.

5. Escrow.com pays the seller after all conditions of the transaction are met.

Many online shopping sites (such as Amazon.com) also sell HDTV-ready sets bearing the familiar computermakers' logos, so don't be surprised if you run across those brands in those venues. However, we recommend that you always visit the manufacturer's home site, so you can see what special deals and bundles it's offering.

What about Amazon.com, eBay.com, and Cnet.com?

When you think about buying anything online, Amazon.com and eBay.com certainly come to mind. For those of us with a more technical slant, Cnet.com comes to mind as well. These sites have expanded their horizons immensely from their book, auction, and technical roots, to encompass a huge legion of sellers of all sorts of goods — including HDTVs. With the exception of some models shown on Amazon, these sites mostly serve as price-comparison engines, showing you where you can buy the units displayed for the least cash.

Amazon.com has a special section for buying HDTVs, located by navigating to Electronics⇨Categories⇨Audio & Video ⇨TVs & HDTVs⇨HDTVs. There you can see direct deals that Amazon has for selling the items under its own storefront, as well as links to major vendors like Best Buy and Crutchfield. As with most of Amazon's entries, you get detailed information about the product, customer reviews, and (here's what we like) what other items people looked at when they were looking at this product.

As far as eBay is concerned, this is a more unpredictable adventure. We priced one 60-inch plasma on eBay and found pricing for the system — brand new, in the box — ranged from $2,200 to $17,999. It's hard to explain such a price range, but definitely think twice before you send that much money to someone on eBay (or any stranger). Be aware that many of these vendors may not be authorized vendors — and (as we mentioned earlier in the chapter) that could mean a voided warranty. Still, there are valid, great deals to be found on eBay, as with many of the other items sold on the site. Caveat emptor, especially with newer vendors who don't have a long track record.

Cnet (www.cnet.com) has really emerged as a powerful force in comparison shopping as well, with content deals with all the major players to show pricing and inventory on the Cnet site. Cnet does not sell anything itself, making Cnet a truly independent source of information. What's powerful about Cnet is that it offers its own reviews of each product, and reflects users' opinions as well. Cnet.com should be a stop on your HDTV hunt.

DISH Network

DISH Network — the satellite TV service provider — bundles HDTV systems with its satellite receivers as part of an all-in-one service that takes the guesswork out of much of this process. Where DISH shines is in bundling an HDTV system at a very attractive price, albeit one that comes with quality limitations and a requirement to become a DISH subscriber for a monthly fee.

That's not all bad — the bundle enables you to immediately set up an entire system and then tap into HDTV content, but it also can mean (in some areas) setting up a separate antenna system to receive broadcast channels. The DISH system comes with an HDTV receiver and a choice of television sets. Be aware, however, that the company announces a package ahead of its actual availability, so don't plan on acting immediately on the latest offering.

This is also less attractive option if you live beyond the reach of broadcast HD channels; you can't get those channels over this system. Before going in this direction, check out the DISH Network HDTV offerings to make sure they match what you want to watch. Otherwise the savings on the system won't be worth the tradeoffs.

At this writing, other satellite companies offer HDTV receivers and either appropriate satellite receivers or upgrades to existing receivers to get HDTV. Voom (www.voom.com) offers professional installation of its satellite service and either outright purchase or monthly rental of the HDTV receiver, to be paired with an HDTV-ready television set. DIRECTV (www.directv.com) has the same.

Circuit City

Circuit City (www.circuitcity.com) combines online ordering with in-store sales and does a pretty good job on both fronts. Circuit City online boasts a wide range of products, but availability is determined by zip code. These products can be compared (up to five items) in a side-by-side format on the Circuit City site.

One major bonus is that the company offers free delivery and setup, and for an extra fee, even connects other video devices within your home to the HDTV system. The vendor also offers immediate advice on the Web site about which accessories are required for which systems, making that part of the purchase easier as well. Customer ratings are provided along with the listings of the HDTV units, but don't seem to vary too much from set to set.

More detailed discussion of HDTV is found in its Learning Center on HDTV, found on the main Televisions page (at Home⇨Televisions). Here you can find an HDTV FAQ, glossary, and a bunch of articles about HDTV (which you're less likely to need since you were smart enough to buy *HDTV For Dummies!*).

Circuit City also offers in-stores sales of HDTV units, which presents a perfect opportunity to see in advance what you are buying, if you are close to a Circuit City store. Circuit City only accepts credit card payments online and is sometimes coy with its pricing, asking you to put an item in your shopping cart before divulging the "secret" discount price. On the plus side, you can order online and pick up the system at the nearest Circuit City store and return it there if need be.

Best Buy

Like Circuit City, Best Buy (www.bestbuy.com) also combines features of a brick-and-mortar store and an online service, capable of price and feature comparisons. Best Buy's online interface offers the option of comparing related systems by features and price. Whether you can buy the system online is tied to your Zip code — not all systems are available even in sites close to Best Buy store locations.

Best Buy has an extensive selection of HDTV systems. The company offers free shipping, but it's more narrowly defined than Circuit City's similar option. You must confirm an address that falls within the free shipping area, and the option doesn't cover all items. If a free shipping option isn't designated, the only way to know whether you qualify is to put the item in your shopping cart, enter the appropriate information and see what Best Buy says. You can pick up the item at a Best Buy store and return it there. Best Buy lists the required accessories, and often offers a discount on these items with the HDTV system.

ClubMac

ClubMac (www.clubmac.com) originally opened to sell — guess what — Apple Mac products. It has expanded its mission to a wide variety of other electronics. So don't be fooled by the Mac-dominated home page. The store meets all the baseline characteristics for online selling. It also offers the widest possible selection of delivery and payment options (including C.O.D.), as well as an 800-number for customer service and online order tracking. The company's HDTV selection includes a variety of manufacturers and pricing options, as well as an extensive set of cables and accessories. ClubMac charges a 30-percent restocking fee if you return the HDTV set.

Super Warehouse

As an online vendor, Super Warehouse (www.superwarehouse.com) is rock-solid on site security, customer service, response, site clarity and ease of use, promptness of order processing and fulfillment, and on-time delivery. It also has an exceptional variety of HDTV choices, in many price ranges. The vendor allows for 20-day product review and doesn't charge a restocking fee, but won't pay for return shipping. Payment options include the usual credit cards, purchase orders, and checks.

Best Digital Online

Best Digital Online meets the table-stakes requirements for an online retailer, providing a secure site that is reasonably easy to navigate, clearly stated policies for customers and well-expressed product descriptions, strong customer service and quick problem resolution and prompt order delivery. It is one of the merchants that charges a restocking fee, although 15 percent is as low as those fees go. It has a wide range of both delivery and payment options, and a wide assortment of both HDTV systems and required accessories. Best Digital Online also provides live customer support representatives to walk you through the purchase process.

Microsoft

With the convergence of the PC and entertainment, Microsoft is not going to be left out of the action. Microsoft's online catalog (www.microsoft.com/windows/catalog/default.aspx) includes information about vendors of cards that enable a personal computer to become an HDTV receiver. These receivers connect to the PC via USB ports or through internal cards that convert a PC to a receiver. This is cutting-edge stuff and not for the technically skittish. The PC option isn't extremely costly at this stage, if you can actually find the available products. At the writing, for example, the "Where to Buy" section of the Sasem Web site is under construction.

Sony Electronics

This option is a limited one — the company has large retail stores in New York and San Francisco, and smaller ones in Chicago, Southern California and Boston. But Sony has indicated plans to expand its retail reach into top U.S. markets, and it is specifically focusing on high-end items such as HDTV units for its retail push. This will probably not be the cheapest option.

Chapter 26

Ten (or so) HDTV Accessories

. . . .

In This Chapter

▶ Shaking it up (baby!) with transducers

▶ Wirelessly saving your marriage

▶ Showing your HDTV movies in the backyard

▶ Going Hollywood all the way

. . . .

No HDTV is an island! It should be connected to and surrounded by accessories that make the HDTV but one part (a big part) of the overall viewing experience. From wireless headphones to vibrating chairs, there are all sorts of ways to extend your HDTV experience.

Kick Some Butt (with Transducers)

One of the best scenes in the movies ever, we think, was in *Jurassic Park,* when Dr. Alan Grant (Sam Neill's character) and his entourage encounter the Tyrannosaurus Rex for the first time, in the unsecured open terrain of the park — who can forget the cup of water vibrating to the footfall of the beast? Scary!

In the movie theater, we all felt that through the incredible bass management of the speaker systems. And in most of our home audio systems, the special effects audio channel on the DVD drives some of that through the subwoofer, too.

But none of that compares to what you can get with audio transducers from companies such as Clark Synthesis (www.clark synthesis.com) and The Guitammer Company Inc. (www.the buttkicker.com). With these transducers, you can get one-giant-T.-Rex-step closer to the ultimate HDTV theater!

Transducers are marketed under many names — *bass shakers, tactile transducers,* or as one vendor calls them, *buttkickers.* The units are screwed onto the bottom of your furniture or into the frames of your floorboards. They are better than subwoofer-based effects, because they can be very localized — only your couch shakes, for instance, if they are attached to the couch's underframing. (Your neighbors will be happy about that!)

If you live in a house that has a lot of concrete, stone, and brickwork, you may definitely want to look at adding a transducer to your furniture, because these homebuilding materials do not conduct the bass very well. Even if you have a great subwoofer, you won't get as much low-frequency effect because of the room's materials. Spend the extra dollars on a transducer in these instances.

As with most parts of your HDTV system, you can spend a lot or a little on transducers. One aspect depends on how many you install. You can one, two or three transducers to a couch, and get an increasingly better effect. With three across the bottom of your sofa, you could power the signal for the middle transducer from the LFE/subwoofer out of the processor-receiver, and drive the two side ones from the left and right front channels, respectively. This gives you feedback across the range of the front speakers. If you are watching *Black Hawk Down,* when there's an explosion on the left, you hear the sound from the right channel as well as feel the lower frequencies from the transducer. If a tank rumbles across the screen, it rumbles across the couch as well.

There's a fairly decent price difference between the lower-budget versions, around $60 for decent Aura Bass Shakers (www.aurasound.com), which do a fairly good job, to the higher-end versions — which have more power and precision — such as the Clark Synthesis or Guitammer Buttkicker models, which can run from $200 to $500 (www.smarthome.com).

There is considerable difference among the units as to how high a frequency they support (that is, for how high a frequency they will react to by vibrating). Some products support frequencies up to 800 Hz. We advise that you keep your frequency range on the lower end — between 5 Hz and 200 Hz. Otherwise, it seems the transducer is always rumbling along in the background, and that gets annoying after a while. You can control the frequency sent to your transducer via an equalizer, such as the $120 Audio Source EQ.

If you mount transducers directly to your furniture, you can get rubberized, molded mounts for your chair or couch legs that isolate the noise and vibration to your furniture.

Read the manufacturer's recommendations closely for info on how to power each unit. We recommend you use one amplifier per transducer. Lower-end models might not need anything more than a 20-watt amplifier; higher-end versions might need an amplifier of at least 100 watts. Often, transducers come bundled with an amplifier (active transducers); consider getting one of these if you have questions. Look for amps that have their own volume control so that you can tune the effect relative to the audio level. (Don't worry about whether it's a high-quality amplifier — its use in this application is not high fidelity.)

Motion Simulators

Now shaking things up with Buttkickers and the like is fun, but the rides at Disney World have progressed quite beyond mere shaking your groove thing — you move up, down, left, right, forward, and backward — all the while it shakes you up.

For the vast majority of us — that is, for those who don't have millions to splurge on their own amusement parks with Disney-World-like rides — such an intense experience has been out of our reach. Until the D-Box.

Unlike bass shakers, which provide only vibration or "shaking" in response to the audio track and are in reality merely transducers that vibrate rather than move air, the Odyssee motion simulator from D-Box (www.d-box.com) is a sophisticated motion simulation system that lifts seating and occupants on an X-Y-Z axis (pitch, roll, and yaw) at up to 2Gs (think F-14 at full throttle) of acceleration.

Odyssee provides dramatic motion that is precisely synchronized with on-screen action, which draws in viewers even more by allowing them to accurately experience the accelerations, turns, and jumps that they could previously only imagine. When a car rounds a corner in a 007 chase scene, turn with it. If the *Top Gun* jet fighter suddenly moves into a climb, climb with it.

Here's how it works: The basic system includes an Odyssee Controller and a set of Odyssee Actuators. The controller manages the translation of motion cues from the DVD's content to the motion of the actuators. The controller's microprocessors direct the vertical and horizontal movement of the actuators. Not all movies are supported by the system; codes must be programmed for each specific movie. More than 200 movies are supported now (see the list at www.d-box.com/en/codes/index.html), with more coming each month.

Just about everyone who's tried this system loves it, but the price tag will set you back. An entry-level system runs $15,000, with the manufacturers shooting to drop the cost of the system even more over the next few years.

Power Conditioner

Most people don't know it, but the electrical power in your home fluctuates all over the place. Every time your refrigerator or air conditioner turns on, there's a surge in the power line to compensate. As devices turn on and off in the house (and in the neighborhood), power levels likewise ebb and flow with the current.

The bottom line is that this affects your HDTV gear. As we mention in Chapter 3, you absolutely must protect your HDTV system with some form of surge protection. But surge protection only protects you from major changes in your electrical lines — you still could have issues with the consistency of the current level going to your expensive gear. For example, if your voltage drops, it may actually lower your amplifier's output. Ground issues can cause hums in your audio and lines on your video display.

Consider buying a power conditioner, which improves and stabilizes the AC power for your HDTV system. Power conditioners use various techniques to restore your AC power to a true 60 Hz, 120-volt signal, which can offer better audio and video performance.

Many of the major wiring vendors offer power conditioners. One of the better-known vendors is Monster Cable, which sells some really cool (and cool-looking) Home Theater Power Centers (www.monster cable.com/power). These products provide surge protection, voltage stabilization, and noise filtering. The more expensive ones even have a neat digital voltmeter readout on the front. (We insist on cool readouts on all our gear!) Expect to typically spend from $100 to $500, to potentially upwards of $1,000, for a power conditioner. Protect your system — get one.

DVD Changer Controllers

If you are anywhere near the movie aficionados that we are, you've got hundreds of videotapes and discs littering your home. What used to be a contest for bragging rights with the neighbor over who bought the most discs at Wal-Mart, has grown into a mess.

The DVD changers we discuss in Chapter 11 are great for smaller collections, but what do you do with the larger collections that you want to be able to access centrally from anywhere in your home?

From the same company that makes the best audio server in the industry comes a video option based on the same technologies. ReQuest Multimedia's VideoReQuest (`www.request.com`) enables you to control up to four Sony DVP-CX777ES 400 disc DVD changers through a simple on-HDTV-screen interface. Imagine being able to instantly select from any of 1,600 DVD discs in a collection with a simple, intuitive, on-screen user interface.

You can Access DVD movies by title, genre, MPAA rating, actor, and director. The VideoReQuest provides for automatic discovery of DVD information using Internet lookup, too — so no more pecking at a keyboard entering videodisc titles (what a pain!). Just load the DVD into the system, and the VideoReQuest takes it from there.

What's more, as with the AudioReQuest, you can integrate your VideoReQuest into your home control network so you can start movies from touchscreens located throughout the home. If you have a home integration system (like Crestron or Elan), VideoReQuest can connect via RS-232 or Ethernet interfaces. The system outputs to any VGA, Component, S-Video, and Composite Video ports.

Simple. Functional. Clean. A perfect solution to a growing problem. The VideoReQuest retails for about $2,500. (But we also highly recommend having an AudioReQuest in your audio system as well to store all your CDs. Check those out at `www.request.com`.)

Show HDTV Outdoors

There's nothing like an outdoor HDTV movie. Starry skies, fresh popcorn, a 10-foot-by-20-foot screen, and lots of friends and blankets. It's easy to do, using common items found at Home Depot and your nearest party tent vendor.

In our experience, the issues with HDTV outdoors are less about having a pristine audio and video experience, as they are about being able to show a huge screen with audio everyone can hear — and picking a good movie. The sheer fun far outweighs any issues that are much more noticeable in an enclosed room built specifically to be a theater.

Your outdoor experience requires these basics:

✔ **A video projector that you can take out to the backyard:** Because of the greater amount of ambient light, look for projectors with at least 1,200 lumens, preferably as much as 2,000 lumens. Aside from that specification, we've had great luck with the Toshiba TLP-S70 that sells on the street for around $1,200.

✔ **A video source:** You can use either a portable DVD player or a wireless link back to your HDTV system.

✔ **Speakers:** You can just run long speaker cables from your source, or pick up some wireless speakers to go the distance. We've used some off-the-shelf 5.8GHz speakers from Radio Shack that work just fine — it changes models occasionally but usually it has a pair that costs around $200. Soon, you'll be able to get 802.11-standard gear in most major computer stores.

✔ **Screen:** You can make an outdoor video screen with the same stuff that party companies use. You can make a screen for $300 to $500 (everything but the pipe on the following list totals for about $100):

   • About 30 10-foot pieces of 1-inch EMT pipe from a home store

   This pipe usually totals at $200 to $400.

   • A 10-foot-by-20-foot party-tent tarp

   • Party-tent junction points for the corners

   • A bunch of ball bungee cords to mount the tarp

If you're showing video outdoors in the summer, cover your gear with a blanket before the show — the evening dew can damage your electronics.

If you have a lot of outdoor shows, consider buying a projector that's optimized to go outdoors. You can get pretty good projectors for less than $1,500. When it isn't in the backyard, you can project it against a screen in your living room or your office. You can check out the latest models ideal for the outdoor experience — as well as ideas for how to do this — at our outdoor home-theater site at www.digitaldummies.com/projects/outdoor.asp.

Creature Double Feature in 3D

Most of us have rather private experiences with 3D, dating back to drive-in movies when we were kids, or relating to some of the stunning latter-day movies from IMAX. Some people have been on the *T2: 3D* "ride" at the Universal Studios theme park, or the *Honey I Shrunk the Kids* exhibit at EPCOT in Disney World.

Regardless of your experience, it's hardly been a common part of your entertainment regimen.

For years, the most common 3D technique used has been the anaglyph, a means of creating a single image from two color-coded images that are superimposed on one another, giving you the sensation of depth perception. The red/blue cheap cardboard glasses are the anaglyph experience that many of us equate to 3D.

However a smart Canadian firm called Sensio (www.sensio.tv) has taken a different path and created the industry's only high-quality 3-D system. Sensio uses a method wherein glasses with electronic LCD shutters alternate left and right, allowing each eye to view the screen every other sixtieth of a second. Each eye only sees its corresponding image (left eye only sees the left angle images) on the screen. The Sensio S3D-100 system is a base unit that is another component in your system. The base unit is connected between your DVD player and your display device. You put in your special Sensio-encoded DVD in your DVD player. The S3D-100 reads the signal coming from the DVD player and transmits alternative (left and right) progressive image to the screen and also sends a signal over an IR emitter to the wireless IR-driven glasses. The IR has a range of 20 to 30 feet in a room.

You can't use the Sensio box with just any TV, DVD player, or DVD. You need a special DVD that carries the Sensio encoding (there are almost two dozen such movies now). You also need something that can take in a VGA connector — the S3D-100's output is analog RGB via a VGA connector, which is common on most video projectors. Most HDTV sets have component video on board — you can handle these with a VGA-to-component converter such as one from Key Digital (the $150 KD-VTCA3).

You also need a DVD player with component video outputs. The Sensio processor requires an NTSC signal, which means you must set a progressive-scan DVD player's output to *interlace* mode. (If you need to know more about progressive and interlace mode, check out Chapter 21).

You won't be able to use the Sensio with a plasma TV because the plasma screens emit a great deal of infrared, which interferes with the glasses' interpretation of the sync signal from the emitter.

Currently, Sensio is trying to expand its cadre of 3D movies. The 250+ 3D movies have been made since about 1915 were displayed using the anaglyph process, meaning that to watch those, you need a special set of anaglyph glasses to watch them; they are useless on a Sensio system. Sensio is presently talking to the major Hollywood

Studio to convert those movies in the Sensio 3D format. The popularity of some more recent movies — *Ghost of The Abyss (Titanic), Shrek 4D, Spy Kids 3D,* and other 3D movies — gives 3D fans like us hope for the future. At $3000 per box, we think the Sensio system is a must-have HDTV accompaniment.

Since the glasses shutter on and off rapidly, Sensio warns that folks who are prone to seizures should avoid wearing the glasses.

Wireless Headphones

There's nothing like a great action-packed movie such as *The Italian Job* screaming over the HDTV. That is, until your spouse leans over and asks you to *turn it down* in no uncertain terms! Wireless headphones not only help you hear your HDTV as loud as you want it, but they also save marriages, too!

Wireless headphones also give you mobility — you can get up and let the dog out, and not miss any action.

These aren't hard to install or use. Wireless headsets use radio to connect to a base station that's connected to your audio receiver.

There's a huge variance in headphones on the market. Some that your whole ear and others fit in your ear. Some headphones deal with ambient noise and use DSPs to process Dolby Digital 5.1.

Look for Dolby Headphones support in your HDTV system gear. Dolby has done wonders with this technology application. See, no matter how many channels or speakers you have in your HDTV setup, you still have just two ears through which all the audio is processed. Dolby Headphone (www.dolby.com/dolbyheadphone/) tries to reproduce what is arriving at your ears by using digital signal processors to simulate Dolby Digital 5.1 over the two headphone channels. Dolby Headphone offers three listening modes, each based on acoustic measurements of real rooms. DH1, or Studio, is a small, well-damped room appropriate for both movies and music-only recordings. DH2 (Cinema) is a more acoustically "live" room, especially suited for music but also great for movies. DH3 (Hall) is a larger room, more like a concert hall or movie theater. All products equipped with Dolby Headphone include DH2. On those offering two or all three environments, you can easily switch between modes to suit the material and your own preferences. So, if you plan on using headphones a lot, see if your receiver supports Dolby Headphone encoding. We highly recommend it.

What headphones do we use? A lot of the wireless headphones geared for the television are outright lousy. The only decent wireless headphones we've run into are made by Sennheiser (www.sennheiserusa.com), specifically the RS65 ($190) and the RS85 ($250). The RS85 sports a high-performance noise-reduction circuit that gives it better performance at the edge of its operating range. Though pricey, both models work great, and their signals really travel far in a home. What's more, you can get an extra headphone for each of the models, so Mom and Dad can both watch loud movies while the kids sleep (Kill that, Bill!).

 Check out a great online resource/store — HeadRoom, at www.headphone.com — for the latest and greatest in headphones, including some good reviews and information. That's where we shop for 'phones (no one paid us to say that either!).

In Search of That Great Remote Control

There are soooo many remote controls on the market, it's daunting just to talk about a few. Remote controls have evolved a lot since the earliest universal remote controls hit the market, allowing you to control multiple devices from one remote control.

A quick scan of HDTV Web sites today shows all sorts of remote-control options: tiny, large, color, touch-sensitive, voice-controlled, time-controlled, and on and on. You can spend $12 on a great remote, or $4,000. You can keep it confined to your HDTV theater or go whole-home.

Types of remotes

Typically, remotes fall into the following categories, presented here in increasing order of functionality and desirability:

- **Standard/dedicated remotes:** These are the device-specific remotes that come with each device.

- **Brand-based remotes that come with a component:** Brand-based remotes are those that work with all sorts of devices from the same manufacturer.

- **Third-party universal remote controls:** *Universal remote controls* are designed to work with any electronics device by way of on-board code databases.

- **Learning remotes:** Learning remotes can learn codes from your existing remote controls. You simply point your remotes at each other, go through a listing of commands, and the remote codes are transferred from one to the other.

- **Programmable remotes:** Programmable remotes allow you to create *macros,* sequential code combinations that do a lot of things at once.

- **Computer-based remotes:** Computer-based remotes take advantage of the growing number of computing devices around the home to provide remote-control capabilities. This category includes PDAs, Web tablets, portable touchscreens, and PCs.

- **Proprietary systems:** A number of closed-system control devices enable you to integrate control of all your home-theater devices on a single control system. Companies such as Niles Audio (www.nilesaudio.com) and Crestron (www.crestron.com) are renowned for their control systems.

Remote controls can use infrared (IR) or radio frequency (RF) to communicate with their base device(s). Some remote-control systems use either tabletop or wall-mounted touchscreen displays.

Two new classes of remotes include two-way operation and voice control. With two-way operation, higher-end remotes can interact with the controlled unit to determine its *state,* so you can determine whether a unit is on, or if the input is already set to Video1, before changing that state. And with two-way operation, you can check your actions to make sure they were carried out, too.

Voice control enables you to talk to your remote control to effect changes in your HDTV environment. Want the lights dimmed and curtains opened? Why not just say, "Start the show," and a preset sequence of activities begins.

More and more remotes have a *docking cradle* where your remote can get charged up, or access the Internet for revised programming schedules or to update its internal code databases.

Sexy remotes

The more you spend on your entertainment system, the more you'll probably spend on your remote controls.

These are some of the best remotes on the market:

✔ **Pronto** (Philips; $199 to $1,400; www.pronto.philips.com)

Philips has a solid line of remote controls that have defined the leading edge of home-theater remotes in a lot of ways.

The $1,400 fully programmable and learning Pronto Pro TSi6400 wireless home control panel combines the best of two infotainment worlds: home-theater system control and 802.11b wireless broadband Internet access. Turn on your HDTV system, check out the programming guide, and then browse the Web. With an optional IR/radio frequency capability, the unit allows you to control infrared devices in cabinets or in other rooms with the simple addition of a Philips RF receiver.

✔ **Harmony** (Logitech; $199 to $299; www.logitech.com/ harmony)

Harmony has been leading the pack toward intelligent, learning, and programmable remotes. Its $299 Harmony SST-768 Remote Control helps you control different actions, such as listening to a baseball game or watching a video. Harmony's remote has a LCD screen and links to your PC via a USB connection to program the remote to tie together multiple actions at once, in order. So you can turn on the TV, the receiver, and the amplifier, and set them all to the correct settings, with one click. Very nice. Harmony was bought by Logitech in 2004, and we LOVE Logitech's gear.

✔ **weemote** (Fobis; $24.95; www.weemote.com)

This is a remote designed just for kids ages 3 to 8, so you can limit the channels they watch and make them responsible for their own remotes (in other words, keep their paws off yours).

✔ **Your PDA as a remote control:** A couple of solutions are out there. Philips has a ProntoLITE ($19.95) for turning Palm-based PDAs into a universal remote control. There are several low-cost Palm OS and Pocket PC OS programs from small shareware developers, too — check out ZDNet's Mobile Downloads area for the latest products (downloads-zdnet.com.com).

✔ **Touchscreens galore around the house:** Crestron (www. crestron.com) rules the upper end of touchscreen options. Crestron's color touchpad systems are to die for, or at least second mortgage for. Other options include HAI's OmniTouch (www.homeauto.com).

Remote control technology advanced quickly. For remote-control options, visit Remote Central (www.remotecentral.com). It has great reviews and tracks the newest remotes.

Bring Hollywood Home

If you really want to show off your HDTV, you can build out your HDTV setup with all the accoutrements of a true theater:

- **Popcorn machines:** Complete with the swing-down popping bucket, these $500-to-$900 machines give you freshly popped popcorn and that movie-theater smell. Check out www.popcorn popper.com for some cool models.

- **Candy stands:** You can get fully lit concession stands for your home theater, just like the ones found in real theaters. You can also get concession signs and candy bins. Check out places such as www.candyfavorites.com.

- **Personalized DVD intro:** Open up each showing with your own customized one-minute video, just like those in the theaters (your DVD player must be able to play DVD-R). For $200 or so from www.htmarket.com, you can personalize one of six home-theater intros, from an awards-night theme to a classic popcorn theme. How fun is that?

- **Furniture:** From full-sized ticket booths to Roaring '20s-style bar furniture, you can recreate almost any environment in your HDTV theater. The sky's the limit on pricing for these items. A ticket booth costs $2,000.

- **Theater-style roping:** You can get that velour-rope-and-stainless-steel-post look if you want — expect to spend around $300 a post and $100 a rope. Think about being able to make your mother-in-law wait in line to see the movie at your new house!

- **Themed carpet:** You can buy carpeting that's festooned with images of cinema stars, film reels, popcorn, and other themes. You can expect to pay up to $50 per square yard or more (as with any carpeting, this can vary greatly).

- **Themed theater lighting:** Cool wall sconces, tabletop lights, and standing lights add the theater look. Budget about $200 for each of these.

- **Film posters:** Complement your HDTV room with the latest movie posters. These are usually a standard 27-by-40-inch size, and cost about $10 to $20 for current films. Go online at www.allposters.com for current films, or get vintage posters at places such as www.moviemarket.com.

You can find these items and more at places such as www.home theaterdecor.com and www.htmarket.com.

Index